CONTENTS

2 Typhoons devastated this woman's home in Thailand.

The Natural History Museum is grateful to The RTZ Corporation PLC for supporting the production of this book.

INTRODUCTION

On the night of October 16, 1987, a hurricane swept across Southern England: 15 million trees were blown down like so many matchsticks, and 160,000 km of road were blocked. Electricity supplies to millions of homes were cut off.

This was a single sharp reminder, in a normally sheltered part of the world, of the power of natural forces. In other regions such events are commonplace, and people live with the daily reality of the power of wind and water, ice and moving rocks, to transform their environment.

The Earth's surface has seen constant change throughout its history of 4600 million years. Violent explosive activity comes in the form of earthquakes and **tsunamis**, landslides and floods. Infinitesimally slow and small actions, over geological stretches of time, also result in great changes to the landscape.

Two major driving mechanisms together bring about this change: the Sun's radiant energy reaching the Earth's surface, and the heat produced by radioactive decay in the Earth's interior. This book deals primarily with the first of these – the Sun as an engine of change – and investigates the agents that remove and redistribute surface material, which are ultimately driven by the input of solar energy. Gravity also plays a key role: potential energy arising from the gravitational attraction of the Earth fuels the down-slope flow of water, ice, **rock** and soil. Under these two forces the agents of change attack and erode the Earth's surface, to reduce its elevation and relief.

But the action of surface forces is not sufficient to explain the shaping of the landscape over time. It is also necessary to take account of the internal engine of change which acts in opposition to them, creating and building up the surface. And we need to understand how these two opposing forces interact to shape the Earth's surface over time.

THE INTERNAL ENGINE
The 1960s and 1970s saw a dramatic advance in our understanding of the Earth as the concept of **plate tectonics** was developed and refined. We now know that the Earth's solid crust is not continuous but is constructed of giant moving plates, on some of which sit the continents. (Continental crust may be as much as 70 km thick below the highest mountains, while oceanic crust is rarely more than 6 km thick.) These multi-billion-tonne slabs of crust are in constant motion, driven by **convection currents** in hot rocks at depth in the interior. Where the plates collide, one is drawn down under the other, recycling material into the interior, causing rocks to melt and rise as volcanic **magma** to the surface. Where the plates pull apart, rising magma solidifies to create new sea floor at giant **mid-oceanic ridges**.

Plate movements are responsible for the massive uplift and buckling of the land surface over huge periods of time, as well as for sporadic volcanic eruptions and earthquakes. It is this internal engine of the Earth that creates the land relief which will in turn be acted on by atmospheric forces at the surface.

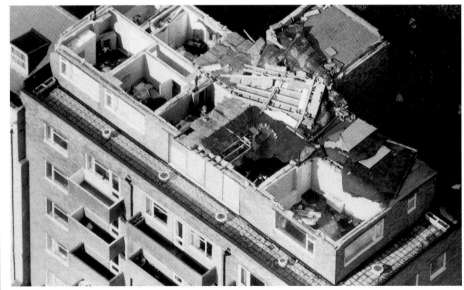

3 Hurricane damage, England 1987.

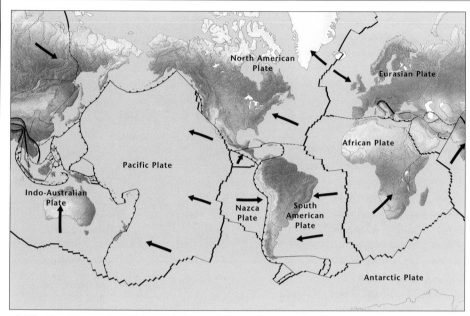

4 The Earth's crust has 12 tectonic plates. Their collision and pulling apart generates unimaginable forces, building mountains and closing or creating oceans.

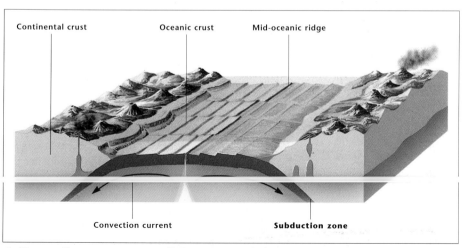

5 New sea-floor crust is continually being generated along mid-oceanic ridges; at continental margins it is pushed down again into the Earth's interior.

STILL MOON

The Moon is our nearest neighbour, so near that we can see some of its surface features with the naked eye. One current theory of its origins has it that a Mars-sized planetary body hit the early Earth with a glancing blow some 4500 million years ago, flinging out the matter to form the Moon.

The Moon's gravity is only one sixth of that on Earth, and so it lacks the gravitational pull to hold onto an atmosphere. It also lacks water. And, while 17% of its surface is covered by ancient lava flows, volcanic activity probably ceased about 3.9–3.1 billion years ago. So, to all intents and purposes, the surface of the Moon is dead. All the forces which mould the Earth's surface are missing.

6 The Moon's surface is heavily pockmarked by the impact of meteorites. Nothing else disturbs it.

WHAT DO YOU MEAN, 'SURFACE'?

In this book, the surface of the Earth is taken to mean not just the solid crust, but also the envelope of gases that protects it, the waters that flow across and through it, and the living organisms that live on it. These four systems interact with one another in complex ways that we are only just beginning to understand. They exchange matter and energy in continuous cycles of change over time. They are in a state of dynamic equilibrium that shifts its balance over geological time.

THE SOLID CRUST

The solid surface of the Earth comprises the soil and the rocks and **minerals** that lie below it on land and beneath the ocean water. Plate tectonics generates new surface rocks from the interior, and processes at the surface reshape these, fragmenting, eroding, transporting and depositing them again to form new rocks.

THE ATMOSPHERE

The atmosphere is a thin blanket of gases, primarily nitrogen and oxygen, held to the Earth by gravitational attraction. Without it, our world would be as lifeless as the Moon.

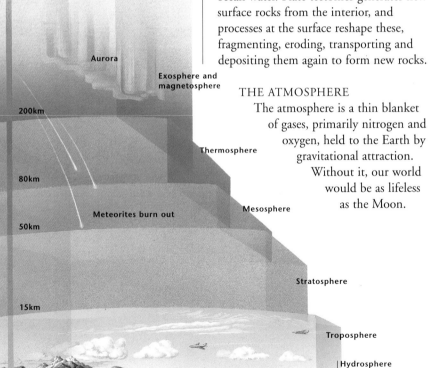

Aurora

Exosphere and magnetosphere

200km

Thermosphere

80km

Meteorites burn out

Mesosphere

50km

Stratosphere

15km

Troposphere

Hydrosphere

Oceanic crust

0

Continental crust

Upper **mantle** – asthenosphere

7

8 *Solid crust – rocks eroded, destroyed and recreated by surface processes.*

The atmosphere keeps us warm, shields us from the Sun's harmful rays and cycles warmth, water and chemicals. It has a complex series of layers stretching some 1000 km out into space, although at this distance very few molecules can be detected. A layer called the **troposphere** contains 99% of the atmospheric gases, concentrated in the first 15 km. Virtually all water vapour and clouds exist in this layer, and almost all weather occurs here. (**Weather** is essentially the atmosphere in turbulent motion.)

By reflecting or absorbing and transporting the Sun's energy, the atmosphere helps the planet balance its energy budget and maintain a global average of air temperature at the surface of about 14°C. The atmosphere allows sunlight through, while gases in the atmosphere such as carbon dioxide and

9 Atmosphere – the thin veil of gases that both protects and moulds the surface.

*10 **Hydrosphere** – coupled with the atmosphere, a key agent of surface change.*

11 Biosphere – all life on Earth, creating and maintaining surface conditions.

water vapour trap heat radiating back from the surface (the **greenhouse effect**, see page 47). Without this barrier, the heat loss would be so rapid that the Earth's surface would cool drastically when not directly warmed by the Sun (the Moon's 'dark side' is minus 120°C). The Earth's atmosphere has evolved with the evolution of life.

THE HYDROSPHERE
Most of the planet is veiled in water, which also makes it unique in our solar system. Water circulates between air, land and ocean. (This pattern of movement, called the hydrologic or **water cycle** is shown on page 38.) Oceans hold 97.3% of the planet's 1.3 billion cubic kilometres of water. A further 2.1% is in ice sheets and **glaciers** – enough for sea levels worldwide to rise an additional 65 m if this ice were

to melt. Groundwater accounts for about 0.6% while 0.01% is in freshwater streams, lakes and rivers. A tiny but critical fraction (0.001%) is in the atmosphere as water vapour. This still amounts to 4000 million tonnes of water, raining on the land and sea every year.

Water currents moderate weather and **climate** by transporting heat from the Equator towards the Poles. And moving water erodes, transports and deposits material to shape the solid surface.

THE BIOSPHERE
The biosphere is the 'layer' at the surface containing all life, in reality spread between the other layers of air, water and earth. Living things are a vital part of the global cycling of materials and energy through the surface systems. They also create the conditions for their own

survival: it was the development of oxygen-releasing organisms that led to the build-up of free oxygen in the atmosphere, on which most life-forms now depend, and which is maintained by living things.

INTERACTING SYSTEM
While each of these four systems can be described separately, it is essential to regard the surface as a single, interacting system. The behaviour of the air and water and of living organisms determines surface processes: flowing water, winds and ocean currents create patterns of **erosion**, transport and deposition to shape the surface. And the form of the surface in turn influences these processes: the position of the continents, the relative position of land and sea, the height and form of mountains all affect the nature of the action of wind and water currents.

DRIVEN BY THE SUN

12 *Energy streaming from the Sun warms the Earth and drives its surface processes.*

WHY IS IT HOTTER AT THE EQUATOR?

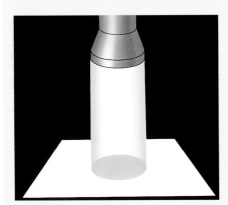

13a *Imagine a torch shining on a flat surface. If the light is held directly over the surface and the beam shines straight down, it illuminates a small area brightly. This is the situation at the Equator.*

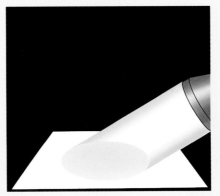

13b *If the torch is held at an angle, a larger area is illuminated, but the intensity of the light is less. One unit of light is spread over a larger area. This is the situation at high latitudes of the globe.*

At the centre of our solar system, the Sun pours out vast quantities of energy into space. We receive only a tiny fraction of this energy on Earth, perhaps only one billionth, but it is sufficient to affect the form and nature of our planet radically. Solar radiation provides the energy for biological activity, the evaporation of water and to drive the global circulation of atmosphere and oceans.

The Sun shines unevenly on the Earth's surface and this simple fact is critical to all that follows. When the Sun shines on the spherical Earth, the Equator receives the most concentrated energy, and is generally warm throughout the year. In winter, northern regions are at such an angle to the Sun that it barely rises, if at all, above the horizon there. Polar regions are cold and ice-bound even in summer.

Because the Sun's radiation falls on the Earth's surface unevenly in this way, it sets in train a restless movement of air and water to distribute this energy across the planet from warm Equator to cold Poles in great convection currents. If air in one region of the globe is heated above the temperature of surrounding air, the warm air becomes less dense and rises. As the warm air rises, cooler, denser air in another portion of the atmosphere sinks. Air flows along the surface to complete the cycle. The steady winds that blow across the tropical oceans are ultimately caused by these churning convection currents.

It might be expected that there would be two giant convection currents, one either side of the Equator, with hot air rising at the Equator and cold air sinking at the Poles – a basic pattern of circulation proposed by the British scientist George Hadley in 1735. In fact, there are six circulatory cells of air motion. Either side of the Equator, air sinks at about 30 degrees latitude, then splits in two and flows north and south from this latitude to create typical patterns of prevailing winds. The winds change direction with the seasons, as different patterns of flow are set up. Their flows determine broad climatic zones worldwide.

Hadley cells are created because the winds are deflected by the spinning of the Earth. Winds flowing from Equator to Pole are deflected in the Northern Hemisphere to the right, and in the Southern Hemisphere to the left. Winds flowing from north and south towards the Equator are deflected in the opposite directions. This deflection is called the **Coriolis effect**.

CLIMATE

Weather is essentially the result of the churning of the atmosphere. And the average of weather over a period of time is known as climate. Because of the global flow of air currents either side of the Equator described above, areas on the same latitude generally experience broadly similar climatic conditions. We can group Earth's major climate zones into just a few major 'belts' classified by temperature and precipitation: polar regions with cold, dry conditions; temperate regions, cool and wet; desert regions, hot and dry; and tropical regions, hot and wet.

However, so many other factors affect regional and local climate that it cannot be predicted by latitude alone; for example, the oceans' ability to retain heat more effectively than air means that coastal climates are generally, although not always, warmer and moister. Mountain ranges also alter the movement of air and generate local climate zones.

Cold, dry conditions often result in a glacial environment, where ice is the major agent of erosion and transport. Hot, dry conditions often result in a desert environment, where wind is the major agent of erosion and transport. And where temperatures and rainfall are both high, in tropical regions, the chemical action on rocks is speeded up, and biological and geological processes are faster.

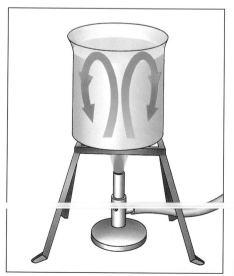

14 *Simple convection current – heated water rises; cooler water sinks.*

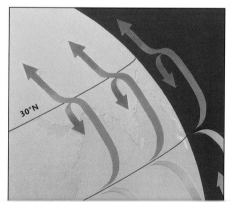

15 *Convection currents in the atmosphere transfer heat away from the Equator.*

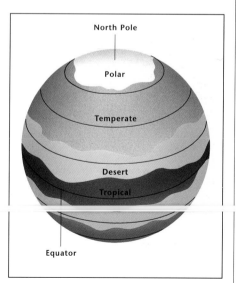

16 *Broad climate zones are created by global air and water currents.*

DRIVEN BY THE SUN

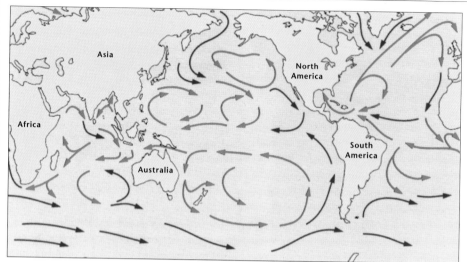

17 *Like giant rivers in the sea, surface currents flow around the oceans. And like the prevailing winds ocean currents are deflected by the spinning of the Earth.*

CURRENTS OF WATER

Ocean currents are driven by the winds of the atmosphere, because when wind blows across water in a constant direction, it drags the water with it. The major global currents move in circular paths called gyres. They too are deflected by the Coriolis effect to rotate clockwise in the Northern Hemisphere, and anti-clockwise in the South. Their pattern of flow is further determined by the arrangement of the continental landmasses.

Like the winds, the oceans transport large quantities of heat northwards and southwards, away from the Equator. The Gulf Stream of the Atlantic, for example, which becomes the North Atlantic Drift before it reaches Europe, carries one

THE SEASONS

The Earth rotates around the Sun once a year. Because it is tilted on its axis, Northern and Southern Hemispheres receive uneven amounts of sunlight at different times of the year. The North Pole tilts towards the Sun in the Northern Hemisphere's summer, and away from it in winter. This means that in summer the Sun is relatively high in the sky, the sunlight is more concentrated and is there for a longer part of the day. (When the North Pole is tilted towards the Sun it receives 24 hours of daylight: the 'midnight sun'.)

This uneven, changing energy input is the cause of the seasons with their changing patterns of winds and currents.

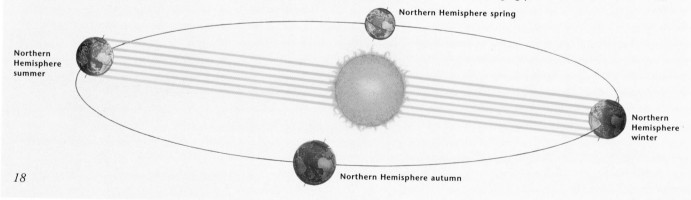

Northern Hemisphere spring

Northern Hemisphere summer

Northern Hemisphere winter

Northern Hemisphere autumn

18

million cubic metres of warm water every second, warming coastal areas. This is why Glasgow in Scotland is relatively mild in comparison with Churchill in Canada, although both are at about the same latitude. (It used to be thought that the deep sea was motionless, unaffected by the surface winds. Now we know that water flows here too. Its movement is driven by differences in salinity and water density: cooled water becomes more dense and sinks, and warmer water rises.)

As well as redistributing the Sun's energy, wind and water also have the ability to erode, transport, and deposit sediments throughout the globe, and are thus the major players in the story of the Earth's restless surface.

19 Coriolis effect – the rotation of the Earth on its axis deflects air and sea currents from their straight paths.

DRIVEN BY THE MOON

The oceans also move in tides, a rhythm of movement in response to the gravitational attraction of the Moon, and to a lesser extent of the Sun. (Although the Moon is much smaller than the Sun, it is so much closer to the Earth that its gravitational influence is stronger.)

At any time, one point of the Earth is directly under the Moon and that part of the ocean is attracted with the strongest force. This attraction causes the ocean to bulge outwards towards the Moon, resulting in a high tide – as the Earth rotates, the bulge always faces the Moon. When both Sun and Moon are directly in line with the Earth, their gravitational fields reinforce each other to create a strong tidal bulge, called a spring tide. When they are 90 degrees out of alignment, each partially offsets the effect of the other and the smallest tidal differences are observed, called neap tides.

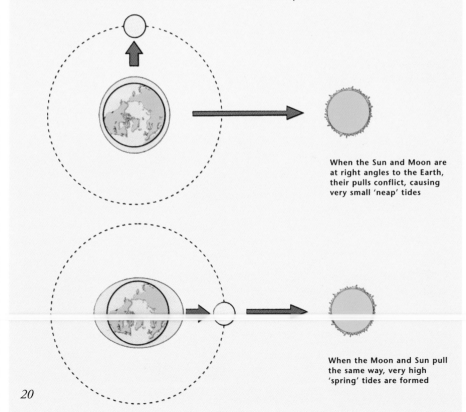

When the Sun and Moon are at right angles to the Earth, their pulls conflict, causing very small 'neap' tides

When the Moon and Sun pull the same way, very high 'spring' tides are formed

20

TIME AND CHANGE

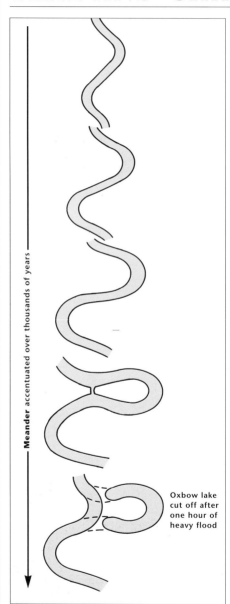

The concept of time is central to geological thought. The processes that shape the surface of the Earth operate over vast expanses of time – millions or even billions of years.

If we discount, for the moment, catastrophic events such as earthquakes and landslides, the Earth's surface seems relatively stable over the timescales we can measure. Historical records do show the slow diversion of a river's course, and the silting up of estuaries, for example, but usually we would expect images taken a hundred years ago to show a landscape essentially the same as today's.

But imagine time speeding up, so that a million years pass in a minute. In this timeframe, we soon lose our concept of 'solid' Earth, as we watch the restless surface change out of all recognition.

There is virtually no place on the Earth's surface that is not moving either vertically or horizontally. The Alps began to rise 40 million years ago, and are still rising today. Scandinavia, too, is rising at the rate of about 1cm per year, rebounding from being weighed down by a vast ice sheet 2–3 km deep which covered it 40,000 years ago. At the same time erosion is also working slowly and inevitably to lower the land. North America's land surface is eroding away at an average of 0.03 mm per year and the Niagara Falls are being cut back by the force of the waterfall at over a metre a year.

Some processes work so slowly that, to our eyes, nothing has changed. The hammering of rain, the molecule-by-molecule dissolution of a rock, the slow scouring of a glacier – all act slowly to

Meander accentuated over thousands of years

Oxbow lake cut off after one hour of heavy flood

21 The path a river cuts over thousands of years can change in a few hours.

22 Colorado River, USA – cutting down the Grand Canyon for over 20 million years.

23 Uluru (Ayers Rock), Australia – Earth's largest monolith, mostly buried in sediment.

wear down the land. Measured on a grand scale, a mountain chain may take ten million years to build, and ten million years to erode.

Many events are not continuous but periodic: spring floods will wash down large quantities of fresh sediment to a lake, then the supply dwindles for the rest of the year. And sometimes change comes with terrifying speed. A sudden flash flood can accomplish in an hour what slow erosion has taken thousands of years to achieve. Over geological time, each of these are important. Each can alter whole landscapes beyond recognition.

CHANGING CONCEPTS

It is only in the last two hundred years or so that we have come to terms with the vastness of geological time. At the end of the 18th century, common belief in the West still held that the Earth had been shaped by catastrophic events as described in the Bible, and most people thought that the Earth was only 6000 years old. But geologists studying rock formations in the field were piecing together a very different story, with a very different timescale.

James Hutton, an Edinburgh geologist of the late 18th century, was among the first to argue that natural agents such as flowing water, acting slowly over thousands or even millions of years, shaped the Earth's surface. Furthermore, he proposed that the surface went through a great cycle of change with 'no vestige of a beginning, no prospect of an end'. This cycle was, he suggested, driven by the erosive power of rivers that wore down mountains and washed down sediment to the sea to form new rocks. Then a great upheaval in the surface thrust those same rocks up once more, and created new (**igneous**) **rocks**, to continue the cycle.

The argument that neither cataclysmic forces nor divine intervention were necessary to explain the structures of the surface was developed further by Charles Lyell, whose 1830 publication, *Principles of Geology*, still guides geological thought. In this, Lyell formulates the principle of **uniformitarianism**. Simply put, it states that in geology we should assume that natural laws are constant in time and space, and that the processes we see in action today are sufficient to explain the past – that 'the present is the key to the past'. (Lyell's theory is more complex and subtle than this, and it is worth reading Stephen Jay Gould's *Time's Arrow, Time's Cycle* for an exploration of both his and Hutton's thoughts.)

Uniformitarianism is still a key idea in geological thought today. As with all explanations in science, it has undergone modification as we have learned more about the Earth's processes. Hutton and Lyell certainly had no knowledge of fundamental processes of plate tectonics, nor of climate change, for example. And geologists today are beginning to recognise the importance of change of many different rates: both catastrophic landslides and slow creeping of land influence the form of the surface. The debate is now focused upon the relative importance of rapid and slow processes over geological time.

24 Sir Charles Lyell, 'father of geology'.

25 James Hutton 'attacks' his critics (1787).

SCALE AND CHANGE

26 Under the microscope, we can study a rock's mineral structure and how it changes.

27 By eye, we can observe the structure of a rock and weathering at its surface.

28 Tracking outwards, we can explore the formation of surface features.

Geology is concerned with processes that act at many different scales, from the structure of minerals to the global budget of whole systems. The question of scale is intimately bound up with the concept of time: small events at a micro-scale can be studied within a limited timeframe; expand the scale, and major changes can only be measured across geological time.

At different scales, causality becomes more or less complex. In a mountain stream, the wearing down of a boulder and its ultimate fate can be explained by a few factors. But the formation of a valley has many and complex causes; the formation of a mountain range has so many that its pattern of erosional development may essentially be unpredictable.

Studied on this wider scale, small variations in the nature and rate of change become insignificant – just as a distant view of a hillside describes its shape but fails to reveal the roughness of its contours. We may observe radical differences in the behaviour of a river from spring flood to high summer drought, and the kind of change it brings about. But take a step back and analyse the river system averaged over thousands of years and these variations will be less significant in our description of how it has changed.

29 On a wider scale, we can study the formation of distinctive landscapes: mountains and valleys, and their evolution over geological time.

The two concepts of scale and time are closely connected.

Modern geological research is increasingly concerned with the large scale and the grand timeframe, as scientists seek to understand the complex dynamics of whole-Earth systems and how they interrelate: how rivers feed the oceans, and how elements are fed back into the system.

This change in perspective has been made possible because of three key factors. First, since the 1960s, plate tectonics has had a major impact on our understanding of the long-term dynamics of landform development. Second, since the 1970s, we have begun to explore the deep-ocean floor, which has also changed our ability to understand rates and scales of change on land as well as underwater. Third, developments in research and in computer modelling techniques are increasing our understanding of climate change. These information sources help us to build a richer picture of Earth's changing nature.

ROCKS AND MINERALS

At micro-scale, rocks can be seen to be composed of smaller units – mineral grains. A mineral is a naturally occurring chemical of particular composition and with a specific 'architectural' arrangement of its constituent atoms. Minerals are made from elements, basic chemicals such as silicon and oxygen (which together make up 75% of the Earth's crust). Some minerals have complex chemical compositions, others are simple. Quartz, for example, is made of silicon and oxygen. Graphite and diamonds contain only carbon, but each has a different atomic architecture. To date about 3600 different minerals have been discovered but most are very rare. Only 30 or so are common at the Earth's surface.

Minerals are the building blocks of rocks. Some rocks contain only one mineral; others include six or more. In coarse-grained rocks, the component minerals are easy to see. Granite, for example, a common rock exposed at the Earth's surface, can be seen to be made essentially of three minerals: quartz, feldspar and mica. Granite is formed by slow crystallisation in the Earth's interior, and the cooling rate during crystallisation determines the size of the crystals.

30 In 'mineral' water there are simply no minerals at all. The water contains a mix of elements in solution (dissociated into ions).

31 This strikingly blue mineral, brought to The Natural History Museum, London in 1996 turned out to be aerinite. Subsequent research on this unusual mineral shows it to have an entirely new kind of structure.

BREAKING UP, BREAKING DOWN

32 *The dramatic tower karst scenery of Guilin, South-West China, forms in steep-sided mountains of metamorphic* **limestone**. *The rock is heavily chemically weathered and eroded in this humid, subtropical environment.*

This chapter is about beginning the process of transformation at the Earth's surface. The surface may seem 'solid as a rock', but over thousands of years even the hardest rocks will be worn away. They are physically and chemically attacked at the surface in a process called **weathering**, just as an iron nail outdoors becomes rusty and fragile and finally turns to dust.

There are two major categories of weathering, although in reality they interact considerably. Chemical weathering is the chemical breakdown, sometimes total dissolution, of rocks; physical weathering is the physical break-up of a rock into smaller fragments.

In both cases, weathering is essentially a process of readjustment of minerals which formed under very different environmental conditions to those prevailing at the Earth's surface. For rocks that were formed deep inside the Earth, the surface is a very different place; some rocks remain very stable on exposure, while others are transformed, or dissolve totally away.

CHEMICAL ATTACK

At the Earth's surface, unlike the interior, there is abundant water and free oxygen, both of which can destabilise minerals. In an oxygen-rich atmosphere, rocks with a high iron content will rust. Oxygen reacts with the iron to form iron oxides (rust), which often remain behind as a red or brown stain in rocks and soils as other elements are removed in solution. Rocks that contain other minerals will be similarly converted to different coloured rusts. For example, copper rust – 'verdigris' – is green.

Given enough time, pure water can dissolve any rock; its ability to do so will depend on the nature of the minerals that make up the rock. But the ability of water to attack certain types of rocks increases immensely if it is slightly acidic. Surface water contains a tiny amount of dissolved carbon dioxide which combines with it to form carbonic acid.

The breakdown of granite is largely due to one chemical reaction – between feldspar and acidified water. Feldspar is a key mineral in the Earth's crust, composed of potassium, sodium, silicon, oxygen and aluminium. Water decomposes feldspar to kaolinite (the basis of China **clay** or kaolin) by removing the potassium and sodium and leaving behind an aluminium silicate. Instead of the crystalline feldspar, weathered granite contains loosely adhering kaolinite, and its structure falls apart, releasing the other mineral grains, mica and quartz. (Quartz is very resistant to weathering, both chemical and physical,

which is why mature sandstones are almost entirely quartz.)

Limestone is a tough, fairly resistant rock, forming major landscapes. However, it will dissolve away totally on contact with acid rainwater. This chemical reaction turns the calcium carbonate of limestone into soluble calcium bicarbonate and releases carbon dioxide gas. In turn, carbon dioxide is removed from the atmosphere by dissolving in rainwater.

34 *Fresh granite, not exposed to the surface, has a clean, tight structure.*

This acid rainwater then dissolves more limestone and the cycle continues. As chemical weathering of limestone accounts for more of the total chemical erosion on land than any other rock, this is of global importance. Weathering of limestone in mountains, such as the Himalayas, is thought to have a significant impact on the carbon dioxide content of the atmosphere, with consequences for global climate (see page 46).

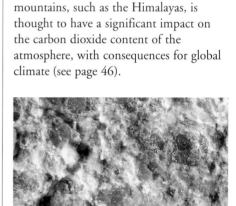

35 *Weathered granite is crumbly and discoloured, altered by chemical attack.*

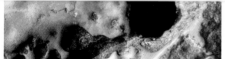

33 *Rock-boring molluscs abrade and dissolve rocks with their acid secretions.*

36 *Many stone buildings are being eroded by acid attack from industrial pollution.*

37 *Lichens growing over a rock surface secrete acids that slowly dissolve away the rock.*

BREAKING UP, BREAKING DOWN

*38 Talus slopes, Svalbard, Norway – loose fragments of rock are the product of **frost-shattering**.*

PHYSICAL WEATHERING

Chemical weathering predominates where temperatures and humidity are sufficient to encourage the chemical reactions that rely on water. But in freezing mountain and polar regions, or in the dry deserts of the world, or where there are few living things, physical disintegration is a much more important factor in determining the weathering of surface rocks.

One of the most efficient physical weathering processes is caused by the rapid freezing and thawing of ice. Water invades the joints or even tiny cracks in a rock, then freezes and thaws, freezes and thaws. The expansion of the water on freezing causes the ice to act like a wedge being driven into the rock, prizing it apart. In high mountain regions and in cold regions the debris of this physical attack lies on slopes as **scree** or **talus**.

The growth of salt crystals in arid areas is thought to have similar effect, expanding as they crystallise to crack rock. This can lead to a form of weathering (in basalts and granites) called **exfoliation**, in which the rock layers spall off from the surface like layers of an onion and may create 'onion-skin boulders'. Because these boulders are typically found in hot, dry lands where temperatures fluctuate dramatically from day to night, the phenomenon was previously thought to be due to the effect of repeated cycles of intense heat and cold, stressing and eventually fracturing the rock. (We do know that violent heat can fracture rock – Hannibal used fire to break boulders on his way across the Alps and, in ancient

BREAKING UP, BREAKING DOWN

Zimbabwe, quarrymen used fire and cold water shock to break rocks.)

Releasing the pressure on rocks can also result in exfoliation. As rocks which formed at great depths become exposed through the erosion of other rocks above, they crack open, rather as though their 'belts' were being loosened and they were allowed to breathe out. Such pressure release causes joints in granite masses, and results in the spalling of thin sheets of rock from boulders and cliffs which can occur on a very large scale.

The more a rock's surface is physically cracked and weakened, the greater is the surface area open to chemical attack. And the more intense the chemical attack, the weaker and more vulnerable the rock is to

physical break-up. Even in the driest desert, some moisture is present and will eventually do its work.

The speed and type of weathering is determined by the parent material, climate, topography, organic activity, and time. Weathering rates vary enormously across climatic zones, and operate on different scales. For example, in temperate regions such as the UK, granites are generally seen as stable, as the climate is relatively cold and the chemical changes proceed slowly. But in the tropics it is a different matter. Here, humidity and high temperatures speed up chemical reactions dramatically, and granite boulders in the ground are so rapidly weathered that they can often easily be kicked to dust.

Weathering is best seen as a process of transformation. When granite weathers to sand and clay, the process of change destroys one form of rock but in doing so creates the materials to make another. Vast quantities of clay are produced to feed the sedimentary process, as is quartz to form sands and sandstones.

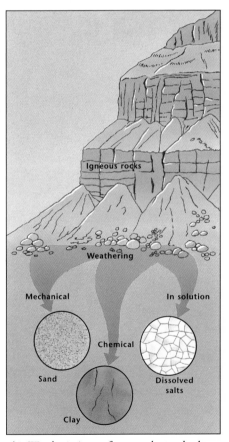

39 *Exfoliated boulder with typical 'onion skin' pattern of weathering.*

40 *Plant power – growing tree roots can exert enough force to crack a rock.*

41 *Weathering transforms – the gradual destruction of granite will eventually yield other rocks.*

SOIL

Perhaps the most important product of weathering for us is soil, which sustains almost all life on land, either directly, in supporting plant life, or indirectly, through the organisms that feed on it. Soil is composed of rock fragments – the products of weathering – and of decaying organic matter – the remains of plants and animals, broken down by insects and trillions of soil-living micro-organisms. Soil is both the product of weathering and the site of further, intense weathering. When soils form, a positive feedback loop exists: the build-up of soil promotes the ideal conditions for further chemical weathering. This is because soil holds and concentrates the water and chemicals that continue the process of decay. Soils may be ten times as acid as rainwater, because they include solutions of carbon dioxide from respiring bacteria and plant roots, and of other organic acids secreted by bacteria. Therefore a covering of soil over rock means that the rock is effectively bathed in acid. So, once soil starts to form, rock weathers more rapidly, and more soil forms.

A single kilogram of soil may be home to an unimaginable number of living organisms. The vast majority are bacteria, fungi and millions of other assorted micro-organisms, and there are thousands of mites, insects, worms, and other creatures. Their activity is vital to the health of the soil. Charles Darwin first highlighted the role of earthworms in turning and aerating the soil, allowing more efficient breakdown and cycling of materials. (He calculated

1 Regolith

Moss and Lichen

Rock fragments

Bedrock

2 Immature soil

A layer of organic material begins to form

Grasses and small shrubs

3 Mature soil

Rotting vegetable matter forms humus

Worm cast

Worms improve the soil texture

Root systems

Burrowing animals break down the soil

Humus

Topsoil

Subsoil

Rock fragments

Parent rock

*42 Soil profile – soils are composed not just of the fertile top zone in which plants grow, but have a complex layered structure. Cutting a pit 1–3 m down into the soil reveals a **soil profile**, with layers called **soil horizons**. These have developed over time, and are the product of decay and the action of water carrying acids and mineral ions down to lower levels.*

43 Soil's raw materials – living matter, gradually broken down, and rock fragments.

that the earthworms in a typical English garden could mix 6.5 tonnes of soil per hectare per year.)

The activity of soil micro-organisms releases elements for use as nutrients by other living things. And chemicals released by weathering may be taken up or transported in solution to be deposited elsewhere, perhaps as cement to bind other rock fragments together, or as crystalline deposits in arid conditions. The process of decay that happens in soil is thus a vital part of the cycle of chemical exchange between living and non-living domains, between water, air, rock and life. Because of this, we should see soil as part of the surface system, its own complex system being part of the Earth's larger interacting systems of exchange.

The way a soil develops depends on many things, but primarily on climate: temperature and rainfall are key factors. Higher rainfall and higher temperatures activate more intense chemical weathering and biological activity. In arid climates, with little water, poor, thin soils develop. In warm, humid climates, deep soils may form, but these may not be as fertile as those of cooler climates. The tropical soils that support the lushest vegetation on Earth are surprisingly unproductive for crop plants. In the heat and humidity, nutrients released from decaying vegetation are quickly recycled in new growth. Also, the heavy rainfall washes rapidly through the soil and carries dissolved minerals down with it in a process called **leaching**. The few minerals which remain are quickly exhausted by

crops, so that in a few years the soil is of no value. Without any input from decaying vegetation it will not be renewed. And without vegetation, it is easily eroded.

VEGETATION ANCHORS THE LAND
Growing plants anchor soil and help to build it up. Sand dunes, for example, build up round nuclei of specially adapted grasses such as marram grass which can survive the hostile conditions. And similarly hardy plants survive and grow in tiny pockets of little more than city grime, gradually accumulating sufficient soil to sustain themselves.

Vegetation is also an important protection against erosion. It reduces windspeeds near the ground and, together with moisture, it tends to bind surface particles together. In arid regions, and over poorly farmed fields stripped of vegetation,

erosion may outstrip the formation of new soil. The infamous 'Dustbowl' of 1930s America was the result of intensive farming during the early 1900s in some south-western states, which damaged and exposed the soil. Intense winds stripped the topsoil, blowing dust more than 1000 km eastwards and ruining over 3.5 million hectares of farming land. Today, millions of tonnes of topsoil are still lost each year in the USA, at twice the rate of soil formation.

Erosion is also a worrying development of increasing tourism in fragile landscapes. Coastal paths, hillsides and even solid rocks are vulnerable. Modern footwear only adds to the problem as increasingly efficient grip disrupts soil cover more effectively. A single walker does little damage, but millions of footsteps put considerable pressure on the system.

44 Thailand – landslide erosion after heavy rain on cleared land.

45 England – a severe case of erosion on a well-trodden path.

EROSION

As weathering slowly weakens and breaks down rocks at the Earth's surface, the more active processes of erosion and transportation take place. Weathering and erosion operate together to lower the surface of the land, and their combined effect is known as **denudation**. This process sculpts the landscape and releases vast quantities of material into the global systems for recycling. Today, we are only beginning to quantify the extent of this process on a global scale.

Gravity is a major source of the power that results in erosion on land. Any material on a slope – rock, soil, water or ice – has **gravitational potential energy**. As it begins to move downwards, this potential energy is transferred into movement – **kinetic energy**. In 'mass movements' such as rock falls and landslides it is the debris produced by weathering itself which moves under the influence of gravity. With rivers and glaciers, gravity provides the energy for water and ice to flow and this flow in turn provides the energy to erode and transport material from the land's surface. When ice and water flow they seek the steepest and shortest route down-slope, ultimately to the sea. This provides the most rapid conversion of gravitational potential energy to kinetic energy.

Water is probably the single most important agent in shaping the form of the land surface. Flowing water has sufficient power to pluck and sweep away soil, grains and pebbles, and even, in flood, to lift and hurl great boulders. Over time, it cuts deep valleys down into the land. And, all the time, it abrades and changes the shape of the surface over which it moves. Waves too can exert enormous power – a wave 10 m high can strike with four times the energy of the three main orbiter engines of the space shuttle.

Ice can also be an important erosional agent. If snow becomes permanent, it becomes compacted and recrystallises to form glacial ice – a crystalline 'rock' made of the mineral ice. As this ice accumulates, it begins to flow under the influence of gravity like a very viscous liquid. The individual molecules of ice slide across one another like cards in a pack, and the whole mass creeps slowly down-slope, bulldozing everything in its path. The huge mass of ice has power to crush and gouge rocks and to carry with it boulders the size of houses. Rock fragments freeze into the base of the glacier and are plucked up and transported. Finer rock debris acts like sandpaper, scratching and scraping the underlying rock.

Wind erosion is a more selective process. It acts effectively when there is no moisture or vegetation to bind loose sediment, and is therefore only a major

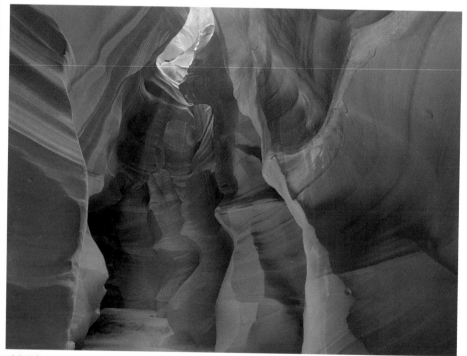

46 The extraordinary beauty of Antelope Canyon, Arizona, USA, owes its form to the erosive action of both wind and water.

erosive agent in dry regions of the world. Because air is much less dense than water, wind can usually move only small rock particles such as sand grains. Larger particles tend to be left behind in a selective process called winnowing. Gradually, these fragments form a gravelly 'desert pavement' protecting the layers of finer material underneath.

The world's hot deserts are a major source of atmospheric dust: an estimated 130–800 million tonnes is blown from the surface of continents annually, most removed in great dust storms. The Sahara has been estimated to lose between 60 million and 200 million tonnes per year. Fine particles can be lifted and carried in suspension in the atmosphere for thousands of miles and for many years. (Studies of **drill cores** from the oceans, where most dust ultimately settles, have helped to establish the sheer volume and effect of windblown erosion on a global scale and the significant contribution of windblown deposits to ocean sediments.)

The other main energy source responsible for erosion and redistribution of material around the globe comes ultimately from the Sun, in driving currents of wind and water, and from the gravitational attraction of both Sun and Moon which together drive the tides (see page 9). The stronger the wind and the greater the expanse of water over which it blows, the larger the waves that buffet the coastlines they encounter. And, as tides rise and fall, water moves in and out from the shore as a broad sheet to erode and redeposit beaches of sand and shingle.

ABRASION

The erosive power of ice, water and wind is greatly increased by the material they carry suspended with them. These particles behave like tools to wear away solid rock.

47

Armed with suspended sand and dust, wind will sandblast rock surfaces, sometimes forming **pedestal rocks**. These may form because a harder, resistant cap of rock lies above softer underlying rock, which has eroded away. Or the structure may mark the upper limit of the sand's abrasive action.

Ventifacts are wind-abraded pebbles created by sandblasting on the side facing into the wind. Occasionally, a sudden gust or storm will turn the pebble, exposing another face to wear. **Dreikanters** are three-sided ventifacts ('Dreikanter' is German for three corners). Their polished surfaces have been abraded by fine dust.

49

Slowly moving ice drags sharp rock fragments with sufficient power to gouge grooves in boulders and hard bedrock. Finer fragments of crushed rock moving with the ice polish surfaces to a high sheen.

48

Loose rocks in a river rub against one another, smoothing and polishing. Sometimes, pebbles caught and swirled round in a depression in the streambed wear the rock to form a pothole.

50

EROSION AND LANDSCAPE

51 Switzerland's Matterhorn – a classic frost-shattered peak eroded by glacial action.

52 The Yellowstone River, USA, cuts a classic V-shaped valley through the mountains of the famous national park.

Erosive forces sculpt the land. The landscapes we now see around us were produced by past erosion, and their forms can indicate the nature of long-past environments. Our interpretation of these forms relies on our knowledge of processes active today – understanding the present is the key to understanding the past.

Glacial erosion leaves very characteristic marks on the landscape. Ice forming at the head of a mountain valley gradually erodes a hollow where the snow accumulates (called a cirque). If two glaciers form on either side of a rock divide they gradually erode that divide away to form a sharp narrow ridge (called an arête). And if three or more are involved, a pinnacle or 'horn' may result, such as the dramatic Matterhorn in Switzerland. Moving glaciers gouge out great **U-shaped valleys** that may deepen enough to cut across valleys of tributary glaciers to leave these as **hanging valleys**.

In contrast to glaciers, rivers cut deep, **V-shaped valleys** through the rock, their natural cutting action being straight downwards, always seeking the fastest, least resistant route downhill. They can form spectacular erosional landscapes such as the eroded valley of the Yellowstone River which cuts through the national park in Wyoming, USA.

Over time, land may be uplifted, or relative sea levels may fall for other reasons, altering a river's gradient. This re-activates the river's erosive action and causes it to cut even further down to the new base level. The Nile, for example, once cut a canyon the size of the Grand Canyon, when the Mediterranean had dried up (see page 39) and the river had to cut much deeper before reaching the sea. The Rhône flowing into the Mediterranean at its western end also cut its own, even bigger canyon. Both of these canyons are now beneath the sea and have filled with sediment, so their presence can be detected only by seismic echo-sounding of the sea floor.

While wind erosion can lead to very distinctive forms, as seen on the previous pages, not all desert landforms can be attributed to wind action. Water can still

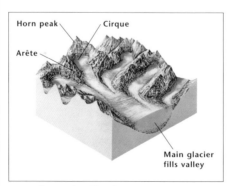

53 Glaciated valley during glaciation.

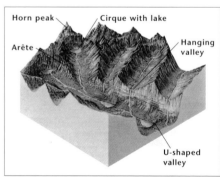

54 Glaciated valley after glaciation.

be a powerful agent in the desert; rare sudden floods can rapidly scour out valleys in the loose sediment, and eroded water-tracts are preserved as **wadis**.

It is at the coast that some of the most dramatic effects of water erosion can be seen, in the destructive power of waves. Waves crashing into a cliff-face gradually undercut its base, until eventually the overhanging rock collapses. Continued erosion causes the cliff-line to retreat leaving a wave-cut platform.

Any weakness in the rock is exploited. Blowholes can form where crashing waves force air under great pressure up through cracks in the rock, and on a larger scale whole coastlines of less resistant rocks are cut back to form bays and inlets. Caves form in more resistant headlands and are quarried through by the sea to form an arch; when an arch collapses it leaves isolated pillars called **stacks**.

55 Stacks, formed by marine erosion.

UNDERWATER AVALANCHES

Erosion is not just a land event: avalanches under the sea create deep submarine canyons much like their land counterparts. These avalanches occur in the vast shelves of loose sediment that build up off the world's great landmasses (see pages 28–29). Occasionally the margins of these shelves collapse, and an underwater avalanche of sediment cascades chaotically downslope to the ocean plains. The water current set up by this sudden movement is called a **turbidity current**. Eventually the current wanes and the sediment it carries gradually settles as a **turbidite** sediment.

57 The Aberystwyth Grits, UK, are huge turbidite deposits which formed 430–440 million years ago. Each layer represents an ancient underwater avalanche event.

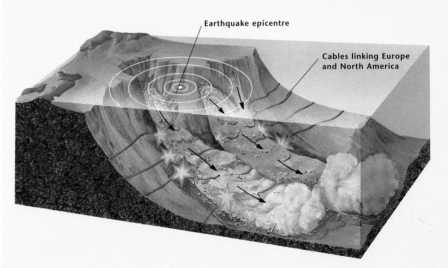

56 Turbidity currents are usually triggered by a sudden erosive slump. One dramatic example of the force and speed of a turbidity current came in 1929, triggered by an earthquake off the Grand Banks of Newfoundland, USA. About an hour after the earthquake, a series of underwater telegraph cables snapped and could be tracked and timed. The current carried sediment for at least 700 km and reached speeds of 40–55 km/hour.

SLOPES ON THE MOVE

The loose debris of weathering and erosion is far less stable than solid rock. Study any slope – whether high mountainside, sea cliff or gentle hill – and you will see the action of gravity: of fallen rocks, in fast mudflow and slow soil creep, and in the sudden slumping of whole tracts of land.

Mass wasting is the movement of material downslope under the influence of gravity – classically without the aid of water, ice or wind, although these may be involved in some way. In order for mass wasting to occur, a slope must become unstable. Like a car parked on a hill, when the hand-brake is released (friction is overcome) the potential energy of the car is converted into kinetic energy – movement. And the steeper the slope the greater the tendency to slide.

The tendency of material to move downhill is dependent on many factors other than the original steepness of the slope: the types, shapes and orientation of the rock fragments or layers, the nature of unconsolidated materials, the presence of water and/or vegetation and the action of earthquakes and volcanoes in triggering the mass-wasting event.

Water can have a dramatic effect on the stability of a slope. First, because water-saturated material is considerably heavier than when dry, it has greater potential energy under gravity. Water also lubricates, reducing friction. Finally, as water freezes and thaws, it alternately pushes apart and re-arranges the packing of particles, weakening their cohesion. Often, the action of water goes unnoticed, until the final collapse. This was the case both in the tragic events at Aberfan in 1966 (see opposite) and in the natural landslide of Bindon, Dorset, England in 1839.

The impact of raindrops alone can erode an exposed sandy slope. Heavy rainfall can cut deep gullies into unstable slopes, or have more catastrophic effects. The water may form a slurry of mud and rocks (mudflow) that can sometimes travel in excess of 120 km/hour.

When a whole hillside becomes unstable and moves, a mass of land or rock may **slump** or fall in one sudden movement. Such slides are usually small-scale, in geological terms. However, there are spectacular exceptions. The largest landslide on Earth, in Saidmarreh, south-west Iran, happened 10,000 years ago. A mass of limestone 15 km long, 5 km wide and 300 m thick travelled 18 km down-slope. The slide was caused by destabilisation of underlying sediments.

Loose material can move downhill as a viscous fluid mass, like road tar on a hot day. Such flows can be extremely rapid, but are usually very gradual. 'Creep' is the slow downhill movement of rock or soil, moving so slowly that it is hard to imagine it as fluid. Its action can be detected in ledges on hillsides. Because the surface material moves more rapidly than deeper levels, objects in the soil are tilted forwards. If you see a tree leaning on a steep slope or tilted gravestones in a hillside graveyard, this generally means that the soil is on the move.

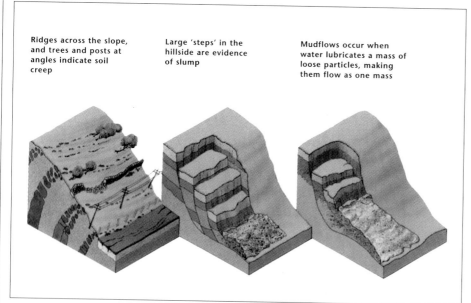

Ridges across the slope, and trees and posts at angles indicate soil creep

Large 'steps' in the hillside are evidence of slump

Mudflows occur when water lubricates a mass of loose particles, making them flow as one mass

58 Basic patterns of movement downslope: soil creep, slump and mudflow.

LANDSLIDES

Eroded material may sometimes flow, slide or fall with astonishing speed, and with terrifying impact on human life. We can't predict the unpredictable. But if we fail to take proper account of how material behaves on a slope, we may court disaster.

59 ABERFAN, 1966, 144 PEOPLE DEAD
After heavy rain, a coal-mine spoil heap 67 m high slid and flowed down-slope onto the Welsh village below. A school and 20 houses were overwhelmed, and 144 people were killed, most of them children.

60 ITALIAN ALPS, 1963, 26,000 DEAD
250 million cubic metres of rock slid into a reservoir high in the Italian Alps. A giant spill-over wave killed 26,000 people downstream.

61 MOUNT ST HELENS, 1980 *The violent eruption of Mount St Helens, USA, blew away the northern flank of the volcano and devastated an area of over 600 km². This final event was triggered by a landslide, which released pressure on the side of the volcano.*

62 ARMERO, 1985, 20,000 PEOPLE DEAD
A volcanic eruption melted snow and ice on this Colombian mountainside. A vast mudflow of water, ash and debris raced down gullies and stream valleys to the town of Armero, 48 km away; 20,000 people were killed.

63 SCARBOROUGH, 1993, HOTEL FALLS OFF CLIFF *Holbeck Hall finally gave in to the waves as the cliff face it stood on collapsed, undercut by wave action which triggered this landslide.*

TRANSPORT

Across the world, vast quantities of matter are travelling, carried in the flow of ice, water and wind. Current global denudation rates, as measured by suspended sediment loads in rivers, are about 15,000 million tonnes a year. Dissolved material accounts for an estimated extra 3700 million tonnes per year. Further millions of tonnes of dust are carried in global wind currents, sometimes for thousands of kilometres. (Dust from central Asia has been found 11,000 km away on the island of Hawaii.)

Each medium – ice, wind, water – has a different capacity to carry material, and this depends on many factors: the speed and steadiness of the flow, the viscosity of the medium, the relative density of medium and particle, and the sizes and shapes of the particles themselves.

Compared to feathers and fine dust, for example, air is a viscous fluid and so these things sink only slowly in air. Wind carries them easily, but even with the force of a tornado, it cannot carry a heavy mass. Storm floods and giant waves, by contrast,

can sweep away structures as large as houses, and ice has the power to hold and carry great boulders for miles. (This is how **erratics** occur – see page 30.)

The properties that most strongly influence the flow of these media are density, **viscosity** and velocity. The less dense a medium is in relation to a particle, the harder it is for it to hold the particle suspended against gravity. Viscosity determines the ability of the fluid to flow. It is the result of the interaction of molecules or particles within the fluid: viscosity increases as the friction between molecules increases, preventing them from easily moving past each other. (The simplest way to understand the effect of this resistance to movement is to imagine the difference between running in open air, through water, and 'wading through treacle'.) As fluid velocity increases, there comes a point when sediment transport can occur, overcoming the opposing forces of gravity and cohesion.

Particles can move along with a fluid in three different ways: rolling, bouncing or

64 Perthshire, UK – flash flooding transforms a quiet river into a raging torrent.

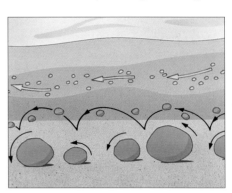

65 Particle movements in water.

swept up into the flow. For any given velocity, where the heavier particles are rolled along, lighter particles move down-current in hops (a process called **saltation**) and the lightest particles are borne along in suspension by the current.

Wind usually lifts only the lightest grains and dust. Turbulent air currents over deserts may transport fine sand particles, while coarser sand particles are transported by bouncing, rolling and sliding along the surface. Heavier grains may saltate along the ground, quickly falling back down to earth under gravity, while pebbles and rocks are left behind.

Water plunges downhill in the youthful upper reaches of a river, abrading and transporting its load of pebbles. Moving more sedately on the gentler, lower slopes it meanders across open valleys carrying with it suspended silt and sand. The turbulence of water lifts and holds small grains in the flow, while larger grains and pebbles tend to be dragged along the bed, moving by small hops.

Rivers swollen in flood are commonly very cloudy with sediment, while slow-moving streams can be crystal clear, their sediment having all settled out.

Ice is so viscous in its flow that it can be considered a solid, but one that still has capacity to flow. Material incorporated into the base and body of the ice is given a 'free ride', held in suspension in the ice until melting releases it. Huge boulders can be carried as easily as fine grains. On melting, the water released from the glacial ice continues to transport some of the finer material, forming 'outwash' deposits.

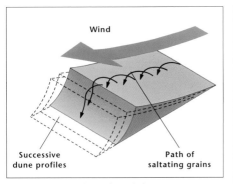

66 *On the surface of sand dunes, grains are usually on the move, causing the whole dune to migrate slowly forward. The wind drives grains of sand to the top of the dune, where they build up and then tumble over the top.*

67 *A dust storm lifts and carries millions of tonnes of sand and finer particles.*

CHANGED BY TRANSPORT

Once rock material is eroded, this is not the end of its transformation – whether it is cobbles, small pebbles or sandgrains buffeted in the wind. Environmental conditions and length of transport both affect continuing changes. Resistant minerals such as quartz are less likely to be destroyed by this abrasion, so they become predominant in the eroded load as it becomes more 'mature'.

Imagine sharp **talus** fragments fall onto a glacier. Frozen into the ice-mass, there is no abrasion between them, and little chemical weathering occurs at such low temperatures. The fragments remain poorly sorted, angular and immature. They will reach the foot of the glacier relatively unchanged.

Imagine next that the talus is swept further downslope by meltwater to join a river. The fragments are rolled and bounced along to round and smooth their jagged edges. Meanwhile, the chemical action of water gradually transforms them, breaking down and dissolving some elements until only the most resistant material is left.

68 *The constant rubbing of rock on rock smooths and rounds particles as they move.*

SETTLING

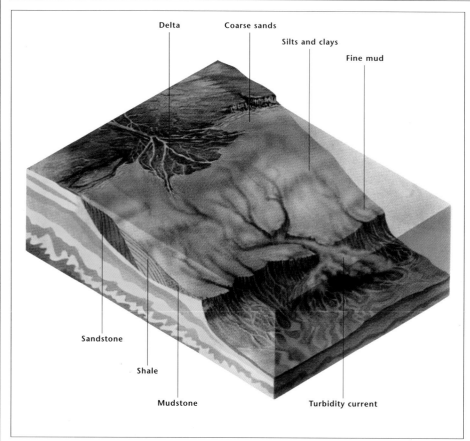

69 *A turbidity current carries sediment under the sea down-slope from a river's mouth.*

Delta   Coarse sands   Silts and clays   Fine mud

Sandstone   Shale   Mudstone   Turbidity current

Worn away and transported by wind, ice and water, rock and mineral fragments eventually settle. Deposition occurs because the transporting medium no longer has the energy to carry the material. Ice melts, winds die down, water flow slows, and their burdens of **sediment** begin to fall under gravity.

Along its course, a river acts like a long-distance sieve, sorting the material it carries as it goes. Wherever it slows, heavier rocks and pebbles are left behind, while finer particles may travel onwards in the flow. Where the river flows into the sea, the far larger volume of seawater dissipates the river's energy and the river sheds its load, again in sequence. Coarser grains drop close to the river mouth, while silts and clays are carried on with the weakening current to settle further out to sea. Most coarse sands will be deposited within tens of metres of the coastline but mud deposits may be found settling up to 30 km out from shore. Sea currents could carry them further.

At the margins of the world's great landmasses lie billions of tonnes of such land-derived sediments. The great majority

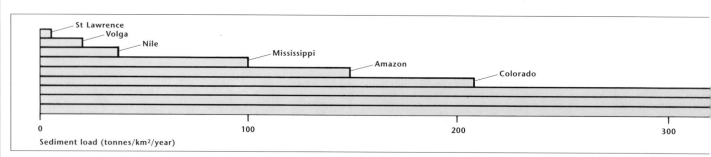

St Lawrence
Volga
Nile
Mississippi
Amazon
Colorado

Sediment load (tonnes/km²/year)

70 *Three great rivers that flow off the Himalayas and Tibetan Plateau carry between them 20% of the world's land-derived sediment.*

of these have been carried down by rivers to the shore, then redistributed by currents in the sea. Some are carried further down the seabed slope by turbidity currents (see page 23). Submarine exploration and the study of deep-sea drill-core samples has made it possible to estimate the vast scale of this transfer of material from land to sea. Submarine fans of turbidite deposits larger than the size of Britain exist off the Mississippi and St Lawrence Rivers. In the Indian Ocean, there is a massive shift of material from continental landmass to seafloor. While erosion removes weight from the continent, the ocean-floor sediments are getting thicker as accumulating sediment weighs down the offshore seabed into the Earth's crust.

Where a river runs into the sea or lake it may deposit sediment in the form of a **delta**. Deltas move gradually as the balance of erosion and deposition alters. As fans of sediment build up and out, the river may abandon its first path and seek other routes to the sea. The Mississippi River has changed its course seven times over the last 6000 years. Its delta, above and below sea level, now extends about 1600 km beyond

the main coastline, forming a 'bird's foot' delta and depositing an estimated two million tonnes of sand and silt a day.

Throughout history, societies have relied on the fertile soils of **floodplains** and deltas to nourish their crops, and learned to live in balance with nature. The people of Bangladesh, for example, live in precarious balance with the rivers that both supply and destroy the land on which they depend. The floods often cause untold damage and misery, but without them Bangladesh would not exist.

Modern developments such as the building of large dams can upset this ancient relationship between people and nature. The Aswan Dam of the Nile, for example, traps vast amounts of sediment. This starves the river of sediment downstream, and is leading to the erosion of the existing fertile delta. Heavy sedimentation also reduces the storage capacity of the reservoir and damages the dam's machinery. Many of the world's large dams are suffering similar problems, caused by human failure to take proper account of the natural processes of transport and deposition.

71 A satellite image of the Mississippi delta, USA, clearly shows its 'birdfoot' form.

72 Bangladesh is a land built from sediment, washed down in great spring floods.

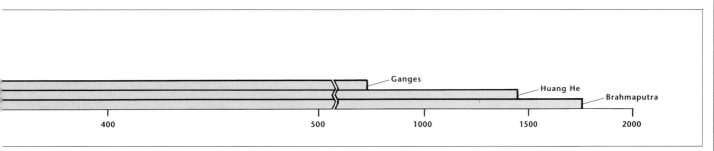

Ganges
Huang He
Brahmaputra

400   500   1000   1500   2000

(Other rivers shown for comparison.)

PLACES TO SETTLE

Sediment settles out when energy levels fall in the transporting medium's flow. This may not always be at the coast; other major sediment traps exist on land – and far out in the deep sea. It may take hundreds or even thousands of years to move this stored sediment elsewhere.

LEFT BY GLACIERS

Besides picking up material by erosion, glaciers also carry fallen rock debris from higher mountains and valley sides. This material is known as **moraine**, and it can be seen as great streaks darkening the glacial ice. When the supporting ice melts, it will be dumped to form a mound or ridge. The deposit itself, called **till**, is an unsorted mix of fragments, from boulders to pebbles and sand. (It was his observations on the close similarities between moraines of modern glacial environments with older deposits and rocks over northern Europe that led the geologist Louis Agassiz to propose in 1840 that the Northern Hemisphere had once been covered by giant ice sheets.)

In areas that have been glaciated, large boulders are occasionally found lying free on rock surfaces from which they were not derived. Boulders of this type are called erratics and may have been transported thousands of kilometres by the moving ice.

As a glacier melts and retreats, streams flow out from within and beneath the ice

74 An erratic.

carrying the finer-grade material with them. This material will gradually be sorted in the water's flow and deposited elsewhere. (See, for example, 'Loess' on the opposite page.)

LAKES

When a river flows into a lake (or artificial reservoir) it will dissipate its energy on entering this large, still volume of water, and sediment settles out. In sufficiently large lakes, deltas may form.

73 A series of terminal moraines deposited by a moving glacier.

75 Lake delta.

ALLUVIAL FANS

When a stream laden with sediment emerges from a steep mountain gorge down to a wide-open plain, it drops its sediment as the velocity and turbulence of the flow decrease rapidly. Gradually a cone of sediment called an **alluvial fan** builds up at the foot of the mountain.

76 Alluvial fan marking break of slope.

FLOODPLAINS

Floodplain deposits arise when a river in a flat-bottomed valley bursts its banks and flows over the adjacent plain, depositing fine sediment over the flat land. These fine silts and clay deposits help produce good, fertile land for agriculture.

77 A river meanders through its floodplain.

DESERTS AND DUNES

In extremely dry areas where there are no rivers to remove sediment and transport it to the sea, sediment accumulates as vast seas of sand known as **ergs**, the classic deserts most people are familiar with. Ergs may cover as much as 500,000 km², as in Saudi Arabia. Sand dunes may be only temporary sediment stores, as the wind continues to pick up and redeposit the sand – and sometimes carry it far away from the desert region. They tend to migrate downwind because sand is constantly moving up the shallow upwind side and being tipped over the steep downwind slope (see figure 66).

LOESS

Vast expanses of accumulated, wind-blown dust and silt called **loess** can be found across the Northern Hemisphere. This sediment probably originated at the margins of retreating ice sheets. High winds blowing off the ice sheet picked up the fine outwash sediment and redeposited it further away as glacial loess. Loess can be carried large distances in suspension. It is estimated that 10% of the world's land surface is covered by loess and loess-like deposits which form extremely fertile soils, for example in China and Eastern Europe.

DEEP SEA

Fine-grained **red clays** are found in the deep parts of the seas. They accumulate very slowly at less than one millimetre every thousand years. These clays may contain fine dust particles that have been blown from the land far out to sea.

78 Sand dunes: an 'erg' desert.

79 Loess deposits.

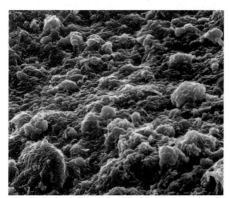

80 Deep sea red clays under an SEM.

LAYERS, BEDS AND OTHER PATTERNS

Beds (layers of sediment) are perhaps the most obvious feature of sedimentary rocks, especially when seen exposed in cross-section at a cliff-face. As deposits of soft sediment build up, they settle in a series of layers. Each is called a bed, and the junctions between them are called **bedding planes**. Series of thin layers, only a few millimetres thick, are called **laminae**, but when the layer is many metres thick, the bed is said to be massive. Slight variations in the composition and structure of the layers means that they may be differentially eroded, to produce a cliff-face that looks like a giant stack of boards.

Sedimentary layers are geological records of the passing of time, and of environmental change. Most of us are familiar with the rings of a deciduous tree trunk, the concentric layers which mark the annual rhythm of summer and winter growth, and there are sediments which also record time in their layers. One remarkable example is the 'Sunday Stone'.

A more natural example is the **varve** layering of sediments, typical of lakes fed by glacial meltwaters. In spring, the meltwater flushes a coarse-grained, light grey clay sediment into the lake, where it settles. Nutrients brought into the lake, and the increasing warmth, promote the growth of microscopic plants called **phytoplankton**. As the season progresses, these create a 'bloom', then die and their microscopic bodies drift to the bottom of the lake to darken the sediment. Year after year, the process is repeated, to create the layers of the varve deposit.

If we make the basic 'uniformitarian' assumption that layers are laid down horizontally and in order, so that the lowest layers are the oldest, then layers can tell a sequential story of change over time. The sequence may be simple, as in the varve, or it may be complex, telling the story of major environmental or climatic changes. The sedimentary layers of the Grand Canyon, for example, record 500 million years of the evolution of life in the fossils preserved in its sequence of layers.

*81 This eroded cliff-face shows sharply defined layers of **sedimentary rock**.*

82 Sunday Stone – a calendar in rock. This sediment formed in a Tyneside coal mine during the 1800s. A white mineral, barium sulphate, settled out in a water trough or conduit and during working shifts it was blackened by coal dust. On Sundays no work was done, and the whole deposit remained white. It is also possible to identify holidays in the sediment's pattern.

83 Varve formation: a spring flush of glacial sediment is followed by a summer bloom of phytoplankton. Dying plankton darken the sediment.

PATTERNS IN THE SEDIMENT

Sedimentary layers often show features that can help in identifying how the sediment was transported and deposited. The most obvious is a sequence of horizontal layers formed by episodes of sediment settling.

Flowing water or wind can sometimes mould the loose sediment into ripples. A current flowing in one direction creates *asymmetrical* ripples, while a tidal current (oscillating backward and forward) forms *symmetrical* ripples.

84

Other, usually transient structures can also be found – marks as delicate as rain spots, the tracks of a skipping stone or the swishing of a seaweed frond across soft mud. Animals that live on and in the sediment may leave their tracks and burrows. Normally, these would disappear as swiftly as a footprint in the sand when the tide comes in. But, very occasionally, new sediment fills the dips and hollows they have made and they are preserved as 'casts' in the hardened rock.

85

In this specimen, a layer of dried-out mud was overlain with a darker sediment, which filled in between the cracks. The whole deposit was then compacted and changed to rock, 'fossilising' the pattern.

A special pattern called **graded bedding** is the result of a sediment of mixed grain-size settling through water. The smaller, lighter and flaky particles encounter more resistance to their fall through water. The heavier particles reach the bottom first, therefore, and the finer grains settle on top of them to form a sequence that fines upwards – a graded bed.

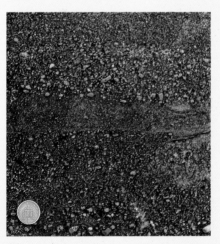

86

READING THE CHANGING PATTERN

87 *Unconformities in rock formations reveal stories of changing environments.*

Sedimentary rocks, being the products of surface processes, are a powerful means of reconstructing the history of the land, and also of understanding the processes and patterns of change. The previous page gave examples of how individual layers and beds can be read to discover how the rocks came to be formed. Series of layers can also record a gradual change over longer periods of time, for example, in the silting up of an estuary to form land.

Sometimes, however, rock sequences preserve evidence of more dramatic changes that took place in the intervening time between the deposition of different beds of sediment. These are called **unconformities**. An unconformity represents a time-gap in the geological record. It marks a distinct 'break' in a sedimentary sequence where deposits have been removed by erosion prior to new sediments being deposited. The 'missing' rocks represented by the unconformity have been eroded away over thousands or even millions of years before the next layer of sediment was laid down.

It was an unconformity that influenced James Hutton to argue for events of long-time change on Earth. One unconformity he studied consists of a series of parallel sedimentary beds lying at an angle beneath a further series of horizontal layers. Hutton was one of the first geologists to realise the implications presented by such field evidence. He proposed that the first layers were sediments laid down in a shallow sea. After being buried deep and hardened to rock, these formations were

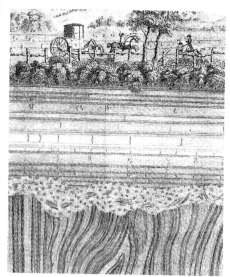

88 *Engraving of the unconformity used to illustrate Hutton's original study.*

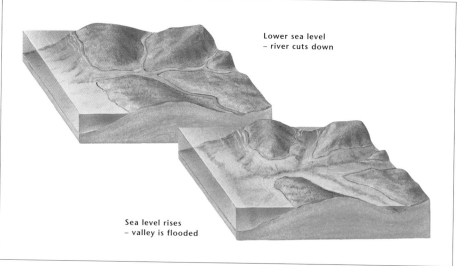

Lower sea level
– river cuts down

Sea level rises
– valley is flooded

89 *Eustasy is the worldwide rising or lowering of sea levels over geological time. It leaves its mark in coastal features such as flooded estuaries and river valleys.*

then tilted, folded and uplifted to the surface, where they were eroded. Much material was removed, levelling these rocks until the sea covered them once more. New sedimentary layers were deposited unconformably on the much older rocks. Then, the layers were thrust up again – to be eroded again. And so, for Hutton, the cycle would continue over immense periods of geological time.

Today, our knowledge of what causes such changes from sedimentary to erosional environments is greatly helped by our knowledge of climate change, and of plate tectonics.

Tectonic movements can thrust rocks upwards from deep levels in the Earth's crust and pile them up into mountains – most spectacularly, today, in the peaks of the Himalayas and Alps. Moving plates also change the pattern and volume of the oceans, and so alter sea levels over geological time. In coastal regions, sedimentary layers may emerge above sea level and become eroded, or submerged below the waves to be overlain with new sediment.

The relative level of land to sea can change dramatically over time because of variations in global climate that affect the amount of water locked up in ice. (Such global change in sea level is called **eustasy**.) In the last **Ice Ages**, so much water was locked up in the ice sheets that the global sea levels fell more than 100 m. Huge deposits of sediment extending from the shoreline on shallow continental shelves were affected most severely by the large-scale changes in sea level associated with

this period of glaciation. They were exposed and eroded, as rivers cut down deep valleys to reach the new, lower sea level. As the seas returned, these valleys were flooded again.

A further effect of the ice sheets was that their added mass weighed down the continental crust, again altering the sea level. Continents sit on a layer of the Earth's mantle called the asthenosphere which is hot and has a plastic, flow-like behaviour. The heavier the continent, the more it sinks into the mantle. Remove the weight and the continent rebounds. (Scandinavia is still rebounding by as much as 2 cm a year from being pressed down by the ice.) This property of readjustment is called **isostasy**.

*91 **Raised beaches** are caused by isostatic readjustment. When ice weighs down the land, a new beach is cut at this level. When the ice melts, the land rebounds and this beach is exposed above the present sea level.*

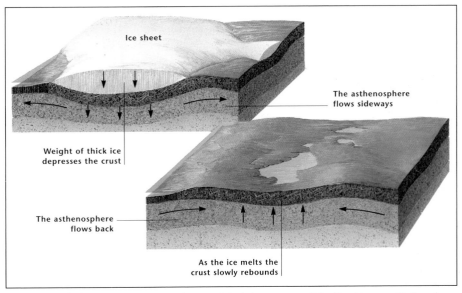

90 During the last Ice Age, vast ice sheets weighed down parts of the Earth's crust. Today, the land is still rebounding. This process of adjustment is called isostasy.

FROM SEDIMENT TO ROCK

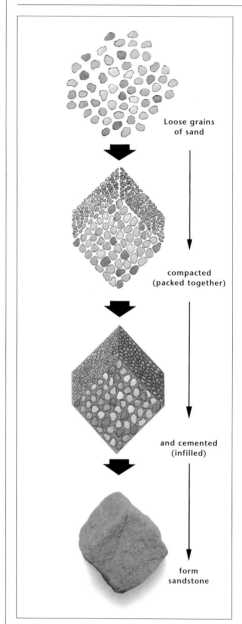

92 From sand to sandstone.

Sediments are constantly shifting, unstable formations. But if they are sheltered from further transport, they may accumulate and become solid rock through a series of physical and chemical changes. Mud and clay will turn to **shale**, sand to sandstone and gravel to conglomerate. The time taken for this to happen varies widely – some still-loose sediments are 30 million years old.

To turn sediment to rock, two main processes must occur: **compaction** and **cementation**. The major physical change is compaction, when the mineral grains of a sediment are squeezed together by the weight of further overlying sediment. Different sediments compact by different amounts. If you look at a sandstone, its structure seems very little different from loose sand, because sand deposits are already fairly well packed. Newly deposited silt and clay deposits, however,

93 Fossil fish, 2 million years old, encapsulated in a concretion.

may contain as much as 90% water and their volume reduces drastically as this water is squeezed out.

The other major processes involved in turning sediment to rock are usually chemical. As water seeps through sediment, minerals dissolved in the water will be deposited as cement to bind the particles of sediment together. The commonest cements are calcite, quartz and iron oxides (the red cement that colours red sandstone). Cementation can also occur when *physical* pressure causes the particles of sediment to dissolve at points of contact. When the dissolved material recrystallises, it will help cement the grains together.

On cementation, further chemical changes may also occur to alter the mineral composition of the rock. In most sediments, the processes of weathering, transport and settling have brought together minerals from different sources. Now, buried and pressed together in a new environment, the mix of minerals gradually readjusts. Some of the original minerals may dissolve and new ones may be precipitated. Occasionally there will be localised changes: minerals may crystallise around a 'seed' such as the remains of an animal or plant, resulting in distinctive concretions that can be easily separated from the surrounding rock (for example flint nodules in **chalk**).

So far, this book has largely been concerned with the erosion, transport and settling of rock fragments to form sedimentary rocks. But there are other sources of material, and other types of sedimentary rocks.

94 The cement that binds the pebbles of this puddingstone is so hard that this rock is ideal for making millstones.

95 Compacted, platy clay particles give shales their typical laminated structure.

96 Biosparite (magnified): shells and other debris bound together by a cement of calcite.

THE ROCK CYCLE

A sedimentary rock is just one stage in a continuing process of change and exchange at the surface of the Earth, and between surface and interior. Hutton proposed, over 200 years ago, that the surface underwent an unending succession of uplift, erosion, sedimentation, burial and melting.

This 'rock cycle' is essentially a grand geochemical cycle, as the planet's minerals are recycled and rearranged to form different rocks. It also forms part of a larger global exchange of material between the solid crust and other surface systems – the atmosphere, hydrosphere and biosphere.

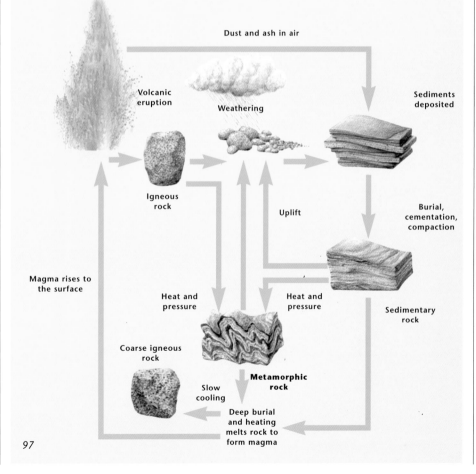

Dust and ash in air

Volcanic eruption

Weathering

Sediments deposited

Igneous rock

Uplift

Burial, cementation, compaction

Magma rises to the surface

Heat and pressure

Heat and pressure

Sedimentary rock

Coarse igneous rock

Metamorphic rock

Slow cooling

Deep burial and heating melts rock to form magma

97

IN AND OUT OF SOLUTION

Natural water is never 'pure': it contains dissolved chemicals that were once part of a rock. Page 15 described how carbon dioxide dissolves in water to make it slightly acidic. This increased acidity increases the water's ability to react with rocks and soils, dissolving the minerals they contain, releasing their constituent atoms into solution as ions.

Water flowing over and through the surface of the land carries away these ions – this is the river's dissolved or **solute** load. The world's rivers carry perhaps 3700 million tonnes of solute a year. The dissolved minerals carried to the sea by the rivers are often characteristic enough for us to detect their 'signature' many kilometres from land. The offshore waters of the Indian Ocean, fed by the rivers running off the Himalayas, have a different chemical composition from those of the Atlantic where the Amazon disgorges. Each river has picked up its own special blend of mineral ions on its journey through different geological terrains.

The sea is saltier than the rivers that feed it, but not as salty as one might expect if this redistribution of minerals ended with its entry into the oceans. Records from ancient deep-sea deposits suggest that the sea's salinity has been relatively stable for much of the Earth's history. Therefore, millions of tonnes of dissolved ions must somehow be removed to maintain this balance. Material is removed by precipitation of crystals out of solution and by incorporation into the bodies of marine organisms. The chemistry of seawater is also profoundly influenced by active cycling of material between ocean and crust. Ocean water is constantly passing through underwater volcanoes at the mid-ocean ridges of spreading plate margins. This water interacts with the hot volcanic rock and extracts chemicals into seawater, and is also responsible for precipitating metal-bearing minerals on the seabed.

Crystallisation that results from evaporation is the most obvious, and easily detectable of these processes. Just as a crust of minerals leached from the soil of a watered pot plant is left behind as the water evaporates, minerals are left behind

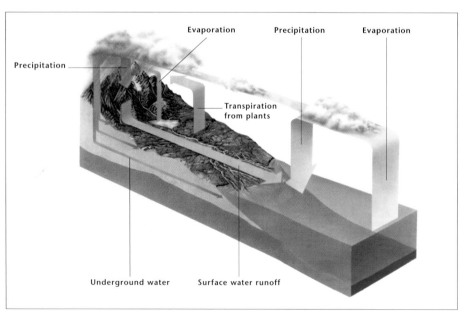

Precipitation

Evaporation   Precipitation   Evaporation

Transpiration from plants

Underground water   Surface water runoff

98 Water cycle – water is constantly recycled between sea, atmosphere and land.

*99 A **'black smoker'**: ocean water heated underground and enriched with minerals is ejected back into the cooler body of water.*

by evaporating seawater. **Evaporite** deposits can form where bodies of seawater become enclosed or temporarily isolated. When seawater is concentrated to about 50% of its original volume by evaporation, minerals begin to crystallise out of solution in reverse order of their solubilities – the least soluble crystallising out first. The chief marine evaporites are gypsum (calcium sulphate hydrate) and halite (sodium chloride) or common salt.

Today the annual rate of evaporation in the Mediterranean Sea is over 4000 km^3 of water, of which only 10% is replenished by rainwater and rivers. The rest still comes from the Atlantic, and if this source were to be cut off, the Mediterranean would dry out. This did happen, six million years ago. The Mediterranean Sea was closed at its western end, and the trapped water – some 4 million km^3 – gradually

evaporated, leaving behind a temporary desert encrusted with a hard layer of evaporites. Five-and-a-half million years ago, the link to the Atlantic was re-established and the sea returned. Gradually new marine sediments buried the evaporite layers hundreds of metres below the seabed.

Seabed soundings of these layers show remarkable structures called salt domes, a few kilometres in diameter and many hundreds of metres thick, penetrating into the overlying sediment beds. Under high pressure salt slowly begins to flow and exhibits 'plastic' behaviour. As it is more buoyant than the overlying sediments, it will flow upwards wherever it finds a weakness, eventually forcing a way through overlying rock strata. Salt domes are economically important because they can create traps for oil and gas to accumulate. By disturbing the rock strata,

they block the flow of these deposits through more porous rock layers.

Seawater is only 3.5% dissolved material, which means that evaporation of even large, isolated seawater masses would not leave more than a thin crust of deposit. But some ancient evaporite deposits are hundreds of metres thick, so the bodies of water from which they crystallised must have been replenished at regular intervals with new saltwater. In North Germany, huge halite deposits 250 million years old are mined today, and there are similar deposits in western Texas and New Mexico. Today's equivalent evaporation environments are the **sabkhas**, coastal flats in the Middle East periodically replenished by the sea, and land-locked salt lakes or **playas** such as those in California's Death Valley, occasionally fed by rainstorms and flash floods (see page 54).

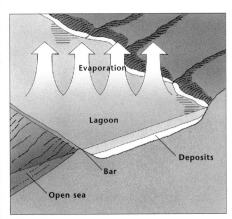

100 Water evaporates from a lagoon, leaving behind an evaporite crust. With occasional flooding, the deposit thickness builds up.

101 A desert basin in Tunisia: in the hot, dry climate, water evaporates to leave crystals of white salt encrusting the plain.

102 Products of evaporation (anti-clockwise from top left): rock salt, satin spar (gypsum) and a **'desert rose'** of concreted sand.

LIMESTONE LANDSCAPES

Limestone can form tough rocks, but will dissolve away completely under attack by acidic water. So in limestone regions, very distinctive geomorphological features and landscapes are created.

Chalk is a relatively soft, pure and porous limestone, with many small joints, but other limestones, such as the Carboniferous limestones of North Yorkshire and South Wales, are harder and denser. These limestones are well-jointed, allowing water to follow a restricted path and not penetrate the rocks as a whole. Typically there is little surface water as most of it sinks down swallow holes and trickles through joints, to re-appear as a stream somewhere at the base of the rock.

On the surface, water erosion may enlarge the joints into a pattern of clefts and ridges, called a **limestone pavement**.

Beneath the ground, the action of water continues. Just as streams erode valleys and create floodplains, underground water etches out the rock to form great underground channels and caves. A cave is fundamentally an erosional feature, but we tend to recall the precipitated deposits that form in them. Where percolating water re-deposits minerals, spectacular structures can develop. Any increase in temperature or decrease in pressure will cause loss of water by evaporation, and result in precipitation of minerals. Excess calcium carbonate is deposited on walls, ceilings

and floors of the caverns creating **dripstone**, **stalactites** and **stalagmites**.

Karst scenery is a term applied to a characteristic weathered limestone landscape where many of the limestone deposits are massive and consist of relatively pure calcium carbonate. (Karst is the German form of a Slovene word for 'bare stony ground'.) The scenery is characterised by a lack of surface drainage, a thin soil, and depressions where surface water sinks into the ground along massive joints, widened by dissolution, and runs underground to form caverns. As these caverns enlarge in time they may eventually collapse creating a hole at the surface. One notable example

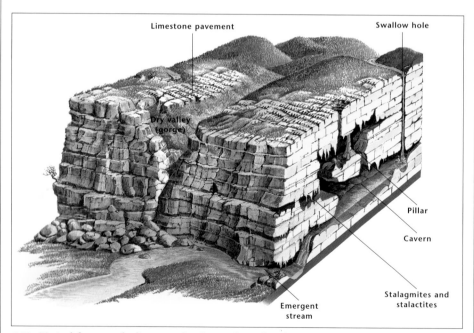

103 Typical features of a limestone landscape, formed as the rock is dissolved by acidified water.

104 Otter Hole, UK – stalactite curtain.

occurred in Florida, in May 1981, when a three-bedroomed house, with half its swimming pool and six cars from a neighbouring parking lot, collapsed into an underground cavern. The hole created eventually broadened to 200 m wide, and 50 m deep.

'Pinnacle' karst, such as the extraordinary sharp-edged topography of Sarawak, is formed in humid tropics by chemical dissolution of metamorphic limestone, with high rainfall, high rates of dissolution and high levels of biological activity. The 'tower' karst of Guilin, China (see page 14), has broad alluvial valley floors and gorges between high towers of limestone undercut by the river water.

105 Needle pillars of eroded pinnacle karst in Sarawak, Malaysia.

106 Mammoth Hot Springs, Yellowstone, USA – terraces of precipitated travertine.

ENDANGERED SPECIES?

Limestone pavement is a horizontal limestone surface, naturally occurring, with deep grooves called 'grikes' between flat-topped areas called 'clints', where the rock has dissolved away along the lines of joints in the bed. Quite how this pattern formed is a matter of debate. Scouring glaciers have exposed the bare rock to acid rainwater attack, although the process of dissolution may have already began while soil still covered the rock. But limestone pavements require further explanation, as many other limestone areas do not show this distinctive topography even though they too were stripped by glacial erosion. The exact way in which massive limestones dissolve is more likely to be related to the cementation structure of the particular limestone.

Limestone pavements take thousands of years to form, and are a unique natural habitat. (The most famous English examples are in the Pennines and around the Lake District. The Burren of Eire is another spectacular example.) Unfortunately, it takes far less time to cut up and remove the rock, which is in great demand for garden rockeries. Few areas of limestone pavement in Britain have been left undamaged, although most remaining areas are now protected. And the strange beauty of these environments also poses a threat to their long-term future – as millions of tourists clamber over them every year, eroding their surfaces further.

107 Malham Cove, Yorkshire, UK – a classic limestone pavement.

LIFE INTO ROCK

Not all rocks at the Earth's surface are formed from the debris of older rocks. Some, such as evaporites (like the travertine illustrated on the previous page), are chemical deposits. And others are made from the debris of long-dead organisms, mostly of marine origin. In the oceans, billions of microscopic creatures depend on the seawater in which they float to supply them with the nutrients and dissolved materials they need. They remove dissolved calcium carbonate from the water to build their bodies, and when they die their remains sink to the sea floor, locking up the minerals they contain in sediments called **oozes**. Shellfish and other larger invertebrates build their shells and other hard parts in the same way and their remains also accumulate in sea-floor sediments.

Two fates can befall such marine sediments: they may be further fragmented and transported elsewhere, for example, on a beach as a drift of shell fragments thrown up by the tide; or, they may be compacted and cemented together where they fall, to form sedimentary rocks. These are the limestones, which vary in purity and composition depending on how and where and from which organisms' remains they were formed. Many contain a significant proportion of sand and clay. Others, such as chalk, are over 80% pure calcium carbonate.

Stromatolites are organic rocks built from mats of cyanobacteria (blue-green algae) or other microscopic algae, which grow as a gluey film over rocks in tidal flats. They build up layer upon layer, each new growth trapping a fine layer of carbonate. Ancient stromatolites are important evidence of the earliest life forms on Earth.

While living organisms store carbon temporarily in their bodies, their

108 Shelly limestone – a fine example of a marine limestone deposit, showing fossilised remains of sea creatures. Sometimes you can see shelly fossils in limestone building materials.

109 Sea bed ooze – the carbon-based shells of microscopic, marine plankton.

accumulated and lithified remains form a longer-term store. Deposits that are made from the remains of living things, rather than the fragments of other rocks, are important components of the Earth's surface, and important 'stores' of carbon. When most of us think about carbon stores we probably think of coal and the other fossil fuels, oil and gas. Although economically the most important examples, these constitute only a very small percentage of carbon deposits on Earth. Most is in stores of limestones such as chalk, ancient coral reefs and the vast oozes of sediment on the ocean floor.

Many ancient limestones have been converted to the mineral **dolomite**, which is a carbonate of magnesium and calcium. It is thought that the limestone has undergone secondary chemical change as magnesium-rich seawater solutions percolated through the existing limestone (calcium carbonate), with the magnesium gradually replacing part of the calcium.

111 The pyramids are built from nummulitic limestone – made up of the hard parts of billions of tiny creatures called **foraminiferans**. *Each single-celled creature builds a spiralling series of chambers of carbonate 'shell' to create a delicate disc. The discs are cemented together into rock.*

112 Reefs build up as corals and **coralline algae** *grow on the skeletons of their predecessors, often on extinct volcanoes. As a volcano gradually sinks below the waves, the corals and algae keep pace by building upwards, as they need sunlight to survive. Many reefs become rock and are preserved.*

110 The Seven Sisters, UK, are made from billions of tiny 'skeletons'. They are the fragmented hard parts, or **coccoliths**, *of microscopic algae that floated in the warm seas that covered this region over 65 million years ago. In death their remains settled to form deep deposits of chalk on the seabed. These white cliffs are an erosional feature exposing a small part of a chalk layer that is 500 m thick.*

THE CARBON CYCLE

Atmospheric carbon dioxide is ultimately the source of all carbon in living things, at the start of the chain that leads from plant photosynthesis to the formation of carbon-based sedimentary rocks. Formation of the rocks removes carbon dioxide from surface reservoirs, to be immobilised in the Earth's crust. Human activities release it much faster than natural processes.

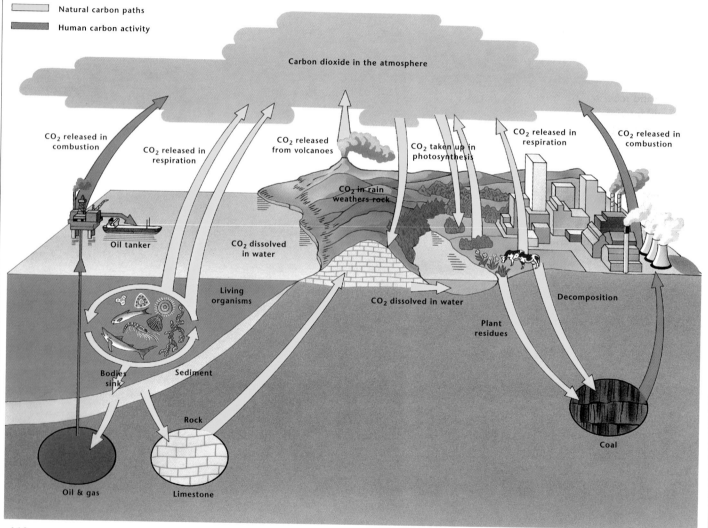

Natural carbon paths
Human carbon activity

Carbon dioxide in the atmosphere

CO_2 released in combustion

CO_2 released in respiration

CO_2 released from volcanoes

CO_2 taken up in photosynthesis

CO_2 released in respiration

CO_2 released in combustion

CO_2 in rain weathers rock

Oil tanker

CO_2 dissolved in water

Living organisms

CO_2 dissolved in water

Decomposition

Plant residues

Bodies sink

Sediment

Rock

Coal

Oil & gas

Limestone

On a global and geological scale, the elements of the Earth's surface are not fixed, but move continually, exchanged between atmosphere, surface and interior. One of the most important of such cycles, for us and all other life on Earth, is that of the element carbon. The carbon cycle is really a summary description of how carbon travels through and between living and non-living realms – between air, water, life and rock, and between the surface and interior of the Earth. At times carbon is captured and stored in a **carbon** sink such as a limestone or coal deposit, then it is released again. Sometimes it may remain locked up for millions of years, sometimes much less. By excavating and burning fossil fuels, we short-circuit the much longer natural cycle to release carbon dioxide back into the atmosphere. This is having profound but as yet unpredictable effects on the planet.

114 Coal sometimes contains concretions called coal balls in which the plant cell structure is preserved unchanged.

COAL

In the presence of oxygen, organic matter will break down and eventually release carbon dioxide back into the atmosphere. But in a low-oxygen environment, such as a stagnant lake, swamp or lagoon, the process of decay will be arrested and hydrocarbons such as coal produced. Coal is a sedimentary rock formed from the remains of trees and other plants under these special conditions which prevented them from simply rotting away.

Coal deposits often occur in a sequence, interleaved with beds of shale and sandstone, which records a rhythmic pattern of past change. Deposition of sand then mud by rivers led to silting-up and formation of a swampy area in which plants could begin to grow. The plant material rotted down to form peat, was buried by incoming mud and compressed, eventually to form coal. The mud created a substrate for renewed plant growth, and the cycle repeated to form the series.

We usually regard coal, and other hydrocarbons such as oil, as sources of energy – or 'fossil fuels' which have trapped the energy of growing living things. Here, our concern is with them as sinks and sources of carbon, particularly in the form of carbon dioxide which is released when the fossil fuels are burned. Carbon dioxide comprises only 0.03% of atmospheric gases, but its concentration has a profound effect on climate. This is examined further on the next pages.

*115 **Coal series** (from top): plant material, peat, lignite, bituminous coal and anthracite. As the coal-forming material is buried and compressed, it undergoes physical and chemical changes. Water and volatile gases escape leaving the deposit richer in carbon. Deeper and longer burial concentrates the carbon.*

CARBON DIOXIDE AND CLIMATE CHANGE

Human activities have been adding carbon dioxide to the atmosphere at an increasing rate since the Industrial Revolution: today, more than five billion tonnes a year are released in burning fossil fuels. Changing agricultural practices worldwide also add **greenhouse gases** to the atmosphere. But is the increase in atmospheric carbon dioxide (CO_2) and other greenhouse gases causing the Earth to warm up? We are now confident that global mean temperatures are rising, with potentially dramatic consequences for **global sea levels**, climatic patterns and for living things. But is there a cause-and-effect link between increasing atmospheric CO_2 and increasing warmth? And what are the implications for the way we live?

By 1988, world concern was growing to such an extent that an International Panel on Climate Change (IPCC) was set up to review current scientific knowledge, assess the environmental and socio-economic impact of climate change, and to formulate response strategies which could be agreed at international level. Atmosphere and climate know no political boundaries and activities in one country or continent can have drastic effects on others, so global action is required.

The IPCC report and other studies estimated an increase in global temperatures in the past century of the order of 0.5°C, and predicted further increases into the next century of about 0.3°C per decade – by the end of the century, the world could be warmer than it has been for a million years. 'Global temperature' means more precisely, the

116 Developing countries such as China are industrialising fast. But the West is still by far the largest emitter of CO_2. (The average Western consumer accounts for 3 tonnes of CO_2 per year.)

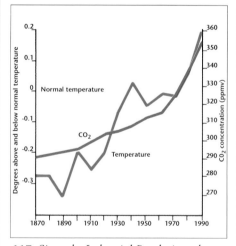

117 Since the Industrial Revolution, the combustion of fossil fuels and deforestation have led to an increase of over 25% in atmospheric CO_2 levels. The rise in global mean temperatures has been less consistent, but an upward trend is now undoubted.

global average annual-mean temperature of the air near the surface of the Earth. Because this is an average, it does not reflect the reality that **global warming**, by affecting the global climate system, may lead to greater extremes of temperatures in some areas (including much colder as well as warmer periods).

There is a general concensus that global warming will lead to a rise in world mean sea levels. There are two components to this rise, polar ice melting and more significantly thermal expansion (as the oceans warm the density of the water decreases and the oceans expand). Predictions range anywhere between 20 cm and 70 cm by the year 2070. There will be major changes in the balance of living communities, which will in turn affect soil conditions, temperature, rainfall, and possibly also food production and population movements.

The reality of the accelerated greenhouse effect is now not disputed, and we know enough to be seriously concerned. But there is no easy answer for what we need to do. We simply do not understand enough about our planet's regulatory mechanisms to be sure. There are so many ways of removing and releasing carbon dioxide from the atmosphere that we cannot predict its behaviour with any certainty. Our uncertainties come from the timing, magnitude and regional patterns of change, and our lack of understanding of which positive and negative feedback systems will come into play. There are also major uncertainties in the behaviour of the oceans and of clouds. Systematic long-term observations are vital to gain a better understanding of climate processes and the role of greenhouse gases.

118 The UK and Eire – not two but a series of islands, if the world's ice melted.

THE GREENHOUSE EFFECT

Carbon dioxide in the atmosphere is essential to life on Earth. Along with other 'greenhouse gases' – such as water vapour, methane, CFCs, ozone and nitrous oxide – it helps balance the Earth's heat budget. Without these gases, the planet would be as inhospitable to life as the Moon.

Incoming solar radiation that is not reflected back into space by the upper layers of the atmosphere is absorbed by the rocks, waters and living organisms of the Earth's surface. In return, the warmed surface re-radiates energy out into space in the form of long-wave infra-red energy (heat). This energy is partially reabsorbed by the greenhouse gases in the air, keeping the Earth warmer than it would otherwise be by some 33°C. This is an entirely natural process, which makes life on Earth possible.

The greenhouse effect is so called because it has been likened to the trapping of warmth by the glass in a greenhouse. The glass acts like the atmosphere by letting in light, but keeping in heat. Global warming is thought to be happening because of human activities adding to the natural greenhouse effect. Predictions suggest that the world may be up to 5°C warmer a hundred years from now.

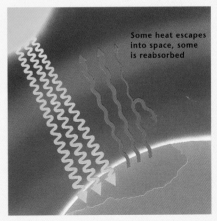

Some heat escapes into space, some is reabsorbed

119

STUDYING CLIMATE CHANGE

In studying climate change one thing is obvious: we are dealing with a highly complex surface system whose nature is inherently changing. A myriad of interlinking factors (apart from the generation of an enhanced greenhouse effect) affect the outcome, and all the surface systems – air, water, rock and life – are intimately involved.

Many feedback mechanisms exist which help to regulate climate: some negative, damping down the change; others positive, accelerating change. The difficulty lies in predicting which will come into operation, and how these mechanisms will interact.

When temperatures rise, for example, the evaporation of water from sea and land increases. The result is an increase in cloud cover, but this could have one of two opposite effects on climate. Cloud cover reflects solar radiation back into space. This is called the **albedo** effect. (Albedo is the reflectivity of the Earth's surface. Snowfields and glaciers have high albedos and reflect 80–90% of sunlight. City buildings and dark pavements have albedos of only 10–20%.) Increased cloud cover could lead to a fall in surface temperatures, as more solar energy is reflected, in a *negative* feedback

mechanism. But increasing cloud cover might also have a *positive* feedback effect, because water vapour is the most important greenhouse gas of all, and an increase could lead to greater retention of heat inside the Earth's atmosphere.

A potential positive feedback loop also exists to enhance the albedo effect when temperatures *fall*. As temperature decreases, ice cover increases. Consequently surface albedo increases, and less heat is absorbed. Surface temperature therefore continues to fall.

As research continues, more and more subtle and complex interactions are discovered which must also be considered to try to understand, and perhaps to predict, climate change. Scientists build highly complex computer models called 'global circulation models', which allow them to undertake simulated experiments on the Earth's response to changes such as the increased burning of fossil fuels. An imaginary atmosphere is modelled, based on data on the known physics of the atmosphere, known patterns of climate change and current understanding of the complex interactions between air, sea, ice and land. Billions of calculations are involved in modelling the climate system and how it might evolve. The details of the model's simulations may not be accurate, but it is more important to recognise any overall changes they predict. These predictions can be compared with data from other sources, such as the fossil record, to see how well they correlate. This process helps to validate the models for use in predicting future climatic change.

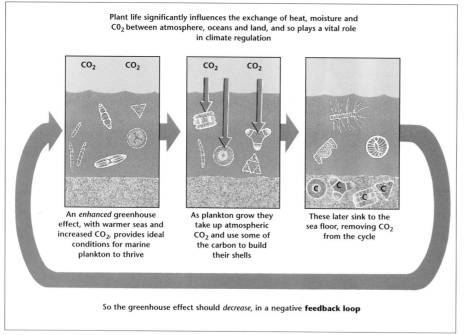

Plant life significantly influences the exchange of heat, moisture and CO_2 between atmosphere, oceans and land, and so plays a vital role in climate regulation

An *enhanced* greenhouse effect, with warmer seas and increased CO_2, provides ideal conditions for marine plankton to thrive

As plankton grow they take up atmospheric CO_2 and use some of the carbon to build their shells

These later sink to the sea floor, removing CO_2 from the cycle

So the greenhouse effect should *decrease*, in a negative **feedback loop**

120 Negative feedback loop: enhanced CO_2 in the atmosphere encourages planktonic growth. In growing, the plankton may eventually absorb the extra CO_2, to restore the balance.

SOURCES OF EVIDENCE

Studying the evidence of past climates is a vital part of learning about the Earth's climate system, and how and why change came about. Many kinds of evidence can be used. For example, pollen and other plant material provide useful indicators of past climates, while fossil corals show growth patterns that record changes in climate and sea level.

More subtly, tiny planktonic fossils also provide a record of past climates. For example, *Globigerina pachyderma* is like a fossil thermometer: in cold water its shell grows by coiling to the left, in warmer water it coils to the right.

Ice cores taken from the Antarctic and Greenland ice caps can also provide a record of past temperatures, as the ice holds 'prehistoric air' trapped in tiny bubbles, which can be extracted and its oxygen analysed. Oxygen exists as two isotopes, O^{16} and O^{18}, and the balance between the amount of these two isotopes dissolved in water varies with temperature: the colder it is, the greater the proportion of O^{18} in the water, the less in air. The air thus has an isotope 'signature' which can provide useful evidence of prehistoric temperatures. A core from Camp County, North Greenland, for example, clearly marks the cold periods of glacial and interglacial phases over the last 150,000 years. Interestingly, this evidence suggests that the climate shifted suddenly rather than gradually between cold snaps and warm spells. This presents a major challenge for the near future; we may be faced with the consequences of our actions more suddenly than we have been expecting.

122 Globigerina pachyderma – a tiny fossil thermometer.

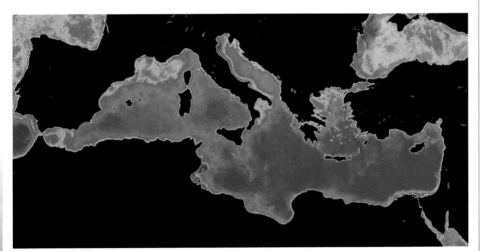

121 Computer-enhanced satellite image shows an algal bloom in the Mediterranean. The growth of such marine plankton takes CO_2 from the atmosphere, but we do not yet know how big a role this plays in regulating global CO_2 levels.

123 Antarctica: a scientist collects an ice core. These cores provide evidence of past climates and climate changes.

COMPLEXITY AND CHANGE

In addition to changes in the composition of the atmosphere, other natural factors appear to alter global climate over time. Changes in solar radiation, tectonic events, slow variations in the Earth's orbit around the Sun, and catastrophic events such as volcanic eruptions and occasional large meteorite impacts, may all be important.

Astronomers have long appreciated that the Sun's output is not constant, and statistical correlations have been made between sunspot frequency and climate change. The nature of the connection (if any) is still under debate, but the onset of the Little Ice Age, a period of extreme cold in 17th and 18th century Europe, has been correlated with contemporary Chinese records of increased sunspot activity.

Changes in the Earth's orbit round the Sun have also been suggested as an explanation for the rapid advances and retreats (reflecting temperature changes) of the last Ice Age. Three types of orbital changes occur:

1 The shape of the Earth's orbit around the Sun changes in a cycle of about 100,000 years.
2 The Earth is tilted on its axis, and this tilt oscillates by 1.5° in a cycle of about 41,000 years.
3 The Earth's axis, which now points directly towards the North Star, wobbles like the axis of a spinning top as it circles the Sun, completing a full cycle every 23,000 years.

In the early 20th century, a Yugoslav scientist called Milutin Milankovitch calculated the combined effects of these cycles and showed how their effect on climate coincided with the known timing of ice age advances and retreats.

The cause of the onset of the Ice Ages is still a mystery, however, although there are many theories. One theory proposes that the raising of the Himalayas and the Tibetan Plateau, with the collision of the Indian and Asian plates, was the trigger.

As you step back in time from the Earth's recent past, it becomes increasingly clear that tectonic movements, driven by interior forces, play a major role in surface change. Further, it becomes important to consider the surface and interior systems as a single, coupled super-system, if we are to try to understand our planet's inherently changing nature.

Tectonic movements are important in influencing local, regional and global climates. The distribution of plates and their continents over the globe, and the relative position of landmasses and oceans, determine the global circulation of wind and water. Evidence from the geological record suggests that the great Ice Ages have occurred during periods when the continents were clustered near to the poles. For example, during the Permian, all the continents were gathered near the South Pole into one giant supercontinent

124 Frost Fair, London, 1613 – during the 'Little Ice Age', the River Thames regularly froze.

called Pangea, and glaciation was extensive at this time. Today, we are in an unusual situation of having not one but two ice caps; in time, these will probably disappear as the continents continue their drifting.

Evidence from studies of the ocean floor (which have been possible only since the 1970s) suggest that the rate and dynamics of sea-floor spreading are also likely to be an important influence on global climate. Sea-floor spreading is, for example, intimately related to the redistribution of carbon dioxide – the CO_2 cycle could be far more dependent on geological activity than on the biological processes which have been previously highlighted. Carbon dioxide is released into the oceans and atmosphere via volcanoes and along the mid-oceanic ridges. On land, chemical weathering of rocks is a major influence in the re-incorporation of CO_2 into rocks.

Such theories are speculative: we are still very much at the beginning of our exploration of the ocean floor. Ocean sediments will continue to provide evidence of past land events, and the study of ocean ridges will continue to yield information about the intricate interchange of water, gases and minerals between interior and surface realms. Each new discovery will continue to add to the complex picture of our world. It becomes increasingly obvious that, such are the complexities of this planet, our understanding of its behaviour will probably always be naive, our knowledge incomplete, our predictions simplistic. But our fascination can only grow.

INDIA HITS ASIA – AN EVENT OF GLOBAL IMPACT

Of all the changes that have occurred on the Earth's surface over the past 50 million years, perhaps the most dramatic has been the collision of the Indian tectonic plate with the Asian plate, and the formation of the Himalayas and Tibetan Plateau. This extraordinary event was itself only part of a constant pattern of change, and it wrought fundamental changes in all the global systems, and, in turn, is shaped by them. The Himalayas are the highest mountains on Earth, yet one day they will be worn away to nothing under the destructive power of surface processes.

As the Himalayas rose, they changed patterns of airflow between land and sea. The monsoons intensified as the growth of a high landmass changed the path of the rain-bearing air. The Gobi Desert formed in an area of once lush ground as the rising Himalayas created a rain-shadow. The intensification of rain over a high relief affected the pattern of weathering dramatically. The Himalayas and Tibetan Plateau are today the most rapidly eroding landscapes in the world; already 6 km^3 have been removed from the land. According to a current theory, this intense weathering drew down carbon dioxide out of the atmosphere so rapidly that it triggered the final dramatic cooling event that led to the Ice Ages.

This change was only one element in a great global rearrangement of oceans and land which also led to the creation of the Atlantic Ocean and the dramatic realignment of global ocean currents. This must have caused major shifts in the distribution of heat and has itself been proposed as a trigger for start of the last Ice Ages. Which theory – or combination of theories – ultimately proves to be correct, will be known only after further scientific investigation and debate.

125 Himalayas: the world's highest mountains.

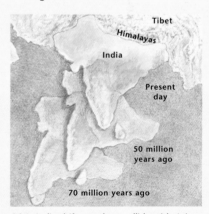

126 India drifts north to collide with Asia.

READING THE LANDSCAPE

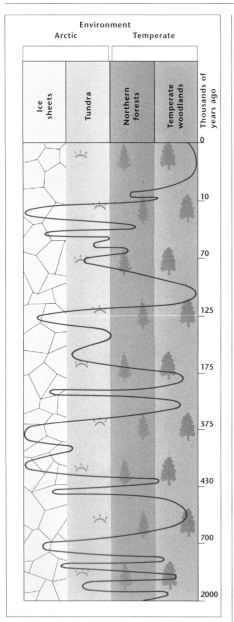

This book has been about changes at the surface of the Earth, and the processes and products of change. It has looked at all scales, from the microscopic to the global. But for us, as human beings, the most important scale is our own – the landscapes we live in and survive on; why they are the way they are, and how they are changing. This last section looks at some typical landscapes, and examines what clues they can give of past environments.

- What types of rocks are present and how did they form?
- What are the dominant erosional landforms and how were they produced?
- What are the processes active today, and what changes will they bring about?

History imposes layer upon layer of change, and in time rocks may be squashed and heated; this masks earlier changes and imposes new patterns. As a result, some of the answers to the above questions may be difficult to disentangle, and advanced geological detective work may be required.

In recent geological time over much of both Northern and Southern Hemispheres, ice has been the dominant sculptor, altering earlier landscapes dramatically. Only a short while ago, ice extended from the polar regions to cover the land, scouring it with ice sheets and glaciers. As water was locked up in ice, land once below sea level was exposed and subjected to the processes of subaerial erosion. The ice

127 Are we heading for another Ice Age?

128 The Tibetan Plateau is one of the fastest-eroding places on Earth today.

sheets have gone through long-term cycles of advance and retreat, and we are now in what was thought to be another 'interglacial' period, which would lead to other ice ages. But, alternatively, there is now the possibility that we might be entering a warmer period, brought about by global warming of our own making. And this will bring different surface processes to bear on the landscape.

Going further back in time over 300 million years, America was joined to what is now Scotland. This ancient continent lay in quite a different part of the globe, subject to a different climate. Scotland was south of the Equator, but England was separated from it by an ocean. As these continental masses both slowly drifted north, they also slowly collided, forming a spectacular belt of mountains in the process. By then, the land had moved sufficiently far north to lie within an arid climate belt, and desert conditions prevailed. Hot desert winds eroded the new mountains, breaking them down into sand that was deposited between the mountain ranges to form sandstones which are being eroded again today.

Later on, as the continent drifted further northwards into tropical climes and dinosaurs lived here, much of the land surface was underwater. There were calm seas in which vast quantities of chalk sediment accumulated and compacted. Later this was uplifted in further tectonic movement, exposing it to erosion before new sediments were unconformably deposited on it. So the rocks of the land formed and reformed, creating new sediments and new rocks out of old, in a never-ending cycle which continues today.

Much of the East Coast of England is now being worn away. Subjected to the battering of the waves, the coastline is changing shape: landslides are common and the coast is retreating some 2 m every year. Since the Domesday Book recorded the settlements of England in 1086, 26 coastal villages have disappeared before the waves. The East Coast's erosion is hastened because it is also slowly sinking, as the tectonic plate which includes Britain re-equilibrates itself after being weighed down into the crust by the last great ice sheets. (The ice weighed more heavily on the West part of the landmass and tipped the East upwards.)

Throughout the world, landscapes record geological change. In many areas of the world not now deserts, you can find the marks of ancient deserts in the rocks. The **cross-bedded** sandstone cliffs in Utah (below) were formed 280 million years ago, when the land lay near the Equator, within a desert climatic belt. Hot winds eroded the older mountains and the evidence of their action can be seen in the rocks. Individual sand grains look like millet seeds, rounded and polished by wind abrasion.

129 Continental drift: Scotland and England joined and then moved north.

130 Norfolk, UK – the eroding and sinking coast has human relevance.

131 Utah, USA – desert-bedded sandstone cliffs re-erode in the hot, dry climate.

A DESERT LANDSCAPE

132 *Desert landscape – sparse vegetation survives between the exposed rocks.*

133 *Badwater Basin, USA – the lowest place on the North American continent.*

134 *Deep erosion by water creates networks of gullies – typical 'badland' features.*

DEATH VALLEY AND
THE MOJAVE DESERT, USA

The high Mojave Desert of California, and Death Valley lying to its north-east, are lands of extremes – some of the hottest, driest places in the Northern Hemisphere. Temperatures can average 49°C in summer. As moisture-laden winds from the Pacific Ocean drop their rain when they rise over mountains near the coastline, the rainfall in the desert is low. There is no soil and little vegetation.

Here one can find an extraordinary range of desert landscapes: sand dunes and flat, gravelly **pediments**; rugged honey-coloured mountain slopes; deeply eroded canyons; and wide valley plains with stagnant marshes and salt flats. These landforms, so obviously shaped by surface processes, owe their existence to major tectonic activities.

Death Valley itself has been called a 'lesson in geology'. The valley was formed some 30–35 million years ago, when complex plate movements stretched and pulled the crust apart in this region of North America. (These same movements formed the notorious San Andreas Fault.) Where the crust was pulled apart, blocks between parallel sets of faults dropped down – in a process called block faulting – to form deep valleys. At its lowest point, at Badwater Basin, Death Valley has the lowest elevation in North America, some 86 m below sea level.

Eroded sediments filled the sinking valley, and intermittent volcanic action triggered by the crustal movements added further sediment. In places, the sediment is so deep that bedrock lies almost 3 km below the surface. The valley has also filled periodically with lakes which then evaporated to leave behind thick layers of salts, which were once mined.

Despite the very low rainfall, water still dominates the shaping of this landscape. Around 200 million years ago seas covered the land, depositing deep layers of sediments that are now thrust up into hills and mountains. Today, water cuts deep into these layers, scouring the hill-slopes of any loose soil and sand and carrying this material down steep-sided wadis to drop it in wide alluvial fans. And the continuing periodic flooding, then evaporation of water, feeds the saltpans which are so characteristic of this landscape.

COASTAL DORSET, ENGLAND –
RETREATING BEFORE THE WAVES

One of the most beautiful coastlines in Britain demonstrates the power of wave erosion. It is characterised by dramatic cliffs, sea arches and stacks that have been formed by wave erosion and are the products of a retreating coastline.

Along the coast around Lyme Regis and Charmouth in Dorset, the cliffs are extremely soft and crumbly, made of a mudrock called **Lias**. This rock formed as soft sediments at the bottom of a shallow sea, 196–203 million years ago. We know that the sea was then teeming with now-extinct creatures such as sea-crocodiles, icthyosaurs, plesiosaurs and ammonites, because their fossilised remains are found in the Lias rocks. As you walk along the shore, especially after heavy rain, you can often see what look like small mudslides where the soft Lias is fluidised to mud again by running water. The delicate cliffs are now in danger not only from the action of waves, rain and landslides, but from people. Year upon year, thousands of visitors come to this part of the coast looking for fossils; many of them accelerate the destruction of the coastline by hammering at the cliffs, even though natural erosion provides more than enough specimens. The best fossils are found by splitting slabs of rocks which have already fallen by natural processes.

Further along the same coast, Lulworth Cove is a small, almost circular bay with a narrow opening to the sea. It is one of Europe's best examples of the action of marine erosion on an alternating sequence of hard and soft layers of sedimentary rocks. The oldest rock is a 150-million-year-old, hard, shelly limestone, which forms a protective outer barrier that can withstand the force of storm waves. Younger beds lain on top of it are of softer material and are more vulnerable to erosion. The beds have been dramatically tilted to near vertical and crumpled into a large S-shaped fold, one of the few examples of the Alpine earth movements in Britain (about 30 million years ago). They dip and plunge away, and have been folded so that the outer line of cliffs is of the hard, older limestones. The sea, however, found a weak point in this barrier and broke through to erode the softer sands and clays behind.

136 Stair Hole: alpine foldings of the outer ridge are gradually succumbing to the waves.

135 Charmouth – the soft Lias cliffs are vulnerable to wave erosion and human attack.

AN ICE-FORMED LANDSCAPE

137 Glencoe – eroded granite peaks.

GLENCOE, SCOTLAND

Some of the most beautiful ice-formed landscapes in Britain are to be found in Wales, Northern Ireland and Scotland, shaped by great ice sheets riding over old mountains on numerous occasions over the past two million years. Ice has far greater powers of erosion than any river: it can fill entire valleys and scoop out rock from valley walls, leaving deep U-shaped valleys such as the dramatic Glencoe in Scotland. This is a landscape still bearing fresh wounds of erosion.

Glencoe is bounded by almost sheer rock faces of ancient granite, with cliffs up to 1000 m high at its pass. Here, the advancing ice breached the mountains, possibly taking advantage of an ancient river valley. The ice cut down an oversteepened valley into weaker metamorphic rocks which lay between huge granitic intrusions. Some of the rocks at the base and the walls of the valleys have been scarred and polished by the ice's abrasive action. The valley has been left littered with the glacial debris carried and dumped by the melting ice. On nearby Rannoch Moor lie some of Britain's most spectacular erratics, originally plucked from the ground in Scandinavia and transported hundreds of kilometres in ice. To the West, Loch Etive is a classic glacial fjord, a submerged glacial valley as deep as the fjords of Norway. Its straight path demonstrates the bulldozing power of the ice that formed it.

138 Rannoch Moor – scoured by ice and now scattered with erratics.

139 Glencoe – the present-day river seems dwarfed in comparison with the deep, wide trough carved by bulldozing ice.

AUSTRALIA'S GREAT BARRIER REEF
The Great Barrier Reef is the largest coral reef on Earth, a living structure covering some 230,000 km². Rather than one reef, it is in fact an intricate series of reefs, stretching more than 2000 km along the eastern coast of Australia and separated from the mainland by a shallow lagoon.

Some of the coral is many millions of years old, but most is much younger. The story of this younger coral is intimately linked with the last Ice Ages, when huge volumes of water were withdrawn from the seas into vast ice sheets. This exposed a wide, shallow continental shelf, thick with sediments, to the east of the Australian mainland. The newly dry land was eroded into hills and valleys, with a range running along the outer rim. Many of the islands on the reef, such as the Whitsunday Group, are the peaks of drowned ridges along this range, now fringed with coral.

By 12,000 years ago, the sea was rising again, flooding the land. Corals began to grow in the shallow water, keeping pace with the rising sea levels and forming the elaborate series of reefs and lagoons we see today. The reef reached its current level about 6000–7000 years ago. Some parts are now well below the growing depth of living coral, extending down some 60 m.

The growth of coral reefs is strongly influenced by the water's warmth, salinity and transparency to sunlight. Reef corals can only grow in water temperatures higher than 17.5°C, and usually no deeper than to about 50 m below the surface. They cannot tolerate low salinities or cloudy waters, and it is the influx of sediment-laden freshwater that determines the northerly limit of the Reef.

142 In the calm waters in the lee of the outer barrier, corals grow in profusion. The main threat to their survival is tourism.

140 The outer reef is unprotected from the powerful waves of the Pacific Ocean, and tropical cyclones can damage it greatly.

141 The Whitsunday Islands are in fact the peaks of drowned mountains.

GLOSSARY

Abrasion The mechanical wearing down of rocks, typically by rubbing or knocking against one another.

Albedo The ability of a surface to reflect the Sun's heat and light.

Alluvial fan A mass of sediment deposited by a river at the base of a steep gradient.

Alluvium A deposit, usually sand and gravel, transported by a river and deposited along its route, mostly on the floodplain.

Bed A layer of sediment. May be thin (see *lamina*) or span several metres.

Bedding plane The surface of rock or sediment parallel to the plane of deposition of a bed.

Black smoker A vent on the sea floor, where heated fluid rich in minerals escapes from the oceanic crust.

Carbon sink Temporary store (although sometimes lasting millions of years) of carbon on or in the surface of the Earth, eg in *limestone*.

Cementation The binding of loose sediment with a cement of material typically deposited round the grains from percolating water.

Chalk A fine-grained white *limestone* composed of microscopic carbonate skeletons of *coccoliths*.

Clay A sediment composed of the finest grade of weathered rock fragments (< 0.002 mm diameter).

Climate The average weather conditions in one area, usually classified by precipitation, temperature and vegetation.

Coal series A series of beds, with seams of coal alternating with shale, mudstones and sandstones, typical of an estuarine swamp environment.

Coccolith Microscopic skeletal plates secreted by plankton, composed of calcium carbonate.

Compaction The slow squashing of sediment under pressure.

Convection current A current in air, water or molten rock, where heated material rises and flows outward from the heat source, and cooler material flows down and inwards to replace it.

Coriolis effect The deflection of currents by the Earth's rotation.

Coralline algae Lime-impregnated red algae.

Cross-bedding A series of inclined bedding planes indicating migration of the bed with the direction of flow of an air or water current.

Delta A fan-like deposit of material where a river flows into a sea or lake.

Denudation Lowering of the land surface, the sum of the surface processes of *erosion* and transportation.

Desert rose Evaporite structure found in desert sandstone, with the appearance of a series of rose petals.

Dolomite A limestone containing more than 15% magnesium.

Dreikanter A three-sided *ventifact* weathered by sand *abrasion*.

Drill-core A sample of rock, or sometimes ice, drilled out to examine its composition and structure.

Dripstone An alternative, general name for *stalagmite*.

Erg Large expanse of sand dune deposits.

Erosion Wearing away and removal of rock (or land) either directly by agents of erosion or by *abrasion*.

Erratic A large boulder transported by ice some distance from its source to an area of different rock type.

Eustasy Worldwide change in sea level, leading to either exposure of seabed or increased covering of land surfaces.

Evaporite Sediment left behind after evaporation of seawater or water from land-locked lakes (see *playa, sabkha*).

Exfoliation A process of weathering in which sheets of rock split off from the surface. Also known as 'onion skin' weathering.

Feedback loop The continuous modification, adjustment or control of a process or system by its own results or effects. Negative feedback damps down change and positive feedback amplifies an effect.

Floodplain The broad expanse or flat land bordering a river, usually in its lower reaches, built up from *alluvium*.

Foraminiferan (plural: foraminifera) Geologically important marine protozoan, normally planktonic. Its carbonate 'shells' form thick layers of marine limestone.

Frost-shattering The weakening and eventual fracture of rocks by repeated cycles of freeze–thaw.

Geochemical Concerned with the chemistry of rocks and the patterns of chemical exchange between rocks, crust and other surface systems.

Glacier A 'river' of slowly moving ice, flowing downhill under gravity.

Global mean sea level The average sea level across the planet (averaged also across tides).

Global warming An increase in global mean temperature. Affected by many factors, the world's temperature has varied widely over its history. Today's global warming is thought to be due to a *greenhouse effect* enhanced by human activity.

Graded bedding A pattern in a *bed*, where the particle sizes become finer towards the top of the bed.

Gravitational potential energy Energy due to the gravitational pull of the Earth, tending to draw all things on the surface down to their lowest level.

Greenhouse effect The trapping of heat radiated from the surface inside the Earth's atmosphere, thus keeping the planet warmer than it would be without an atmosphere.

Greenhouse gas A gas in the atmosphere which reflects back heat radiating from the Earth's surface.

Hanging valley A glacial landform, where the valley of a tributary glacier is cut by the main glacier, which cuts a deeper trough. When the ice retreats the tributary valley is left 'hanging', often forming a waterfall.

Hydrosphere The sum of water on the planet, from atmospheric water vapour to ice and underground reservoirs.

Ice age A cold period in the Earth's history when ice covered much more of the surface. The most recent ice age ended about 10,000 years ago.

Igneous rock One of the three main types of rock (with *metamorphic* and *sedimentary*). Formed from once-molten *magma*, igneous rocks may have a crystalline or 'glassy' structure eg granite, gabbro and basalt.

Isostasy The tendency of the Earth's crust to maintain an equilibrium position in the mantle, eg, slowly rising up after the removal of ice cover.

Karst scenery Landscape showing typical features of limestone erosion, such as sink-holes and caverns.

Kinetic energy Energy due to motion.

Lamina (plural: laminae) A very fine layer of sediment.

Laterite Deep red tropical soil.

Leaching The removal of minerals from soil by dissolution in water.

Lias (Blue) Rocks of Lower Jurassic age, 195-172 million years ago, especially muddy limestone, typically rich in fossils.

Limestone Any sedimentary rock composed primarily of (calcium) carbonate. Most are of biological origin, although some have been chemically precipitated.

Limestone pavement A flat area of exposed, eroded limestone, with deep grooves (grikes) cutting down between

GLOSSARY

flat surfaces (clints).

Loess A fine sediment derived typically from glacial outwash, and further redistributed by wind.

Magma Hot, molten rock from the Earth's *mantle* and lower crust, which crystallises to form *igneous rock*, either at the surface (as lava) or below it.

Mantle The semi-molten interior of the Earth, below the surface crust and above the core. *Convection currents* in the mantle account for the movement of the *tectonic plates*.

Mass wasting The mass-movement of material downslope under gravity, eg rockfalls, landslides, soil creep.

Meander Any bend that develops naturally in a river, gradually exaggerated over time.

Meltwater Water that flows from the foot of a melting glacier (see *outwash*).

Metamorphic rock One of the three basic types of rock (with *igneous* and *sedimentary*), formed by the changing of other existing rocks under heat and pressure, below the surface. For example, limestone can be metamorphosed into marble.

Mid-oceanic ridge A long ridge, like a long mountain range along the ocean floor, where two *tectonic plates* are separating. New sea floor is continually created as molten rock rises up from the Earth's interior.

Mineral A naturally occurring chemical, either a single element or combination of elements, the basic constituent of *rocks*.

Moraine The debris eroded and carried by a *glacier*, eventually deposited when the ice melts (see *till*).

Oolith A tiny sphere of calcite, formed from a 'seed' of shell fragment, rolled to and fro by waves in lime mud of shallow seas. (Many ooliths cemented together form oolitic limestone.)

Ooze A thick deposit on the deep ocean floor, largely composed of the remains of sea organisms such as plankton and shellfish.

Outwash The finer sand and gravel sediment from a glacier, carried onwards in the *meltwater* streams that run from its base.

Pedestal rock A wind-eroded rock found in arid environments, where the base is eroded faster than the top, often forming a mushroom shape.

Pediment A gently inclined plain of eroded bedrock generally veneered with fluvial gravels.

Phytoplankton Aquatic, microscopic plant life-forms.

Plate tectonics The movement of plates of the Earth's crust, pulling apart at some margins, colliding, submerging and crumpling at others.

Playa A land-locked lake which dries periodically, leaving a crust of *evaporites* on the surface.

Raised beach A beach now above the present sea level, formed at a time when the sea level was higher.

Red clay The deepest ocean deposit, formed from fine volcanic ash and dust carried out to sea on the wind.

Rock Any large, solid mass at the Earth's surface, composed of one or more *minerals*.

Rock cycle The continuing transformation and movement of rocks on and in the Earth over time.

Sabkha A salt-encrusted coastal flat, flooded occasionally by the sea.

Saltation The hop-like movement through air or water of particles which are lifted by the flow but are too heavy to stay in suspension.

Scree (or talus) The accumulated frost-shattered debris of rock fragments.

Sediment Loose debris of *weathering* and *erosion*, transported to settle elsewhere.

Sedimentary rock One of the three main rock types (with *igneous* and *metamorphic*), formed from sediment – debris of rocks or organic remains, or precipitating from solution.

SEM Scanning electron microscope.

Shale A sedimentary rock made of fine rock debris, typically flaky.

Sink-hole A feature of limestone scenery, where surface waters go underground.

Slump A type of *mass wasting* in which there is a sudden slippage of sediment as a whole.

Soil horizon A distinct layer seen in a *soil profile*, a region of particular biological and chemical activity.

Soil profile A vertical cross-section of soil from topsoil to bedrock.

Solute Dissolved material.

Stack A rock pillar, free-standing in the sea; remnant of an eroded sea-cliff.

Stalactite A hanging depositional structure, typically of precipitated calcium carbonate on cave roofs.

Stalagmite A rising pillar-like depositional structure, typically of precipitated calcium carbonate and found on the floor of caves (see *dripstone*).

Striations Long scour marks along a boulder or rock-face made by other rocks dragged over it by a moving glacier or ice sheet.

Subduction zone A margin where one *tectonic plate* is dragged under another (the process of subduction).

Talus See *scree*.

Tectonic plate One of 12 segments of the Earth's crust which moves slowly across the planet's surface.

Till A deposit left by a retreating or melting glacier or ice sheet. Typically unsorted, although *meltwater* may carry away the finer material.

Tillite A rock formed from consolidated *till*.

Troposphere The lowest layer of the atmosphere, where major air currents are formed and *weather* occurs.

Tsunami A giant, ocean shock-wave following an earthquake or volcanic eruption.

Turbidite A rock formed from the deposited material of a *turbidity current*.

Turbidity current A cloudy mass of suspended sediment which moves like a current.

U-shaped valley A wide, deep valley scoured by moving ice.

Unconformity A clear break in a sedimentary sequence, which marks a change in the history of that environment, eg when a period of erosion replaced a process of deposition.

Uniformitarianism The fundamental concept that geological processes operate consistently through time, so that knowledge of present processes can be used to explain the past.

V-shaped valley The classic shape of a valley eroded by a river cutting down.

Varve The layer of sediment laid down in a year, as applied to lake sediments which show clear annual cycles.

Ventifact A pebble with facets eroded by wind-sand abrasion.

Viscosity The tendency of material to resist flow ('stickiness').

Wadi A valley carved by an intermittent stream or river in a usually arid environment.

Water cycle The pathways and stores of water circulating on and within the surface of the Earth, including the atmosphere.

Weather The day-to-day conditions in the atmosphere affecting a particular place (see *climate*).

Weathering The physical fragmentation and/or chemical breakdown of rocks at or near the surface of the Earth.

INDEX